T0201444

Understanding Genes

What are genes? What do genes do? These questions are not simple and straightforward to answer; at the same time, simplistic answers are quite prevalent and are taken for granted. This book aims to explain the origin of the gene concept, its various meanings both within and outside science, as well as to debunk the intuitive view of the existence of "genes for" characteristics and disease. Drawing on contemporary research in genetics and genomics, as well as on ideas from history of science, philosophy of science, psychology, and science education, it explains what genes are and what they can and cannot do. By presenting complex concepts and research in a comprehensible and rigorous manner, it examines the potential impact of research in genetics and genomics and how important genes actually are for our lives. *Understanding Genes* is an accessible and engaging introduction to genes for any interested reader.

Kostas Kampourakis is the author and editor of books about evolution, genetics, philosophy, and history of science, and the editor of the Cambridge University Press book series *Understanding Life*. He is a former editor-in-chief of the journal *Science & Education*, and the book series *Science: Philosophy, History and Education*. He is currently a researcher at the University of Geneva, where he also teaches at the Section of Biology and the University Institute for Teacher Education (http://kampourakis.com).

The **Understanding Life Series** is for anyone wanting an engaging and concise way into a key biological topic. Offering a multidisciplinary perspective, these accessible guides address common misconceptions and misunderstandings in a thoughtful way to help stimulate debate and encourage a more in-depth understanding. Written by leading thinkers in each field, these books are for anyone wanting an expert overview that will enable clearer thinking on each topic.

Series Editor: Kostas Kampourakis http://kampourakis.com/

Published titles:

Understanding Evolution	Kostas Kampourakis	9781108746083
Understanding Coronavirus	Raul Rabadan	9781108826716
Understanding Development	Alessandro Minelli	9781108799232
Understanding Evo-Devo	Wallace Arthur	9781108819466
Understanding Genes	Kostas Kampourakis	9781108812825
Understanding DNA Ancestry	Sheldon Krimsky	9781108816038

Forthcoming:

Understanding Intelligence	Ken Richardson	9781108940368
Understanding Metaphors in the Life Sciences	Andrew S. Reynolds	9781108940498
Understanding Creationism	Glenn Branch	9781108927505
Understanding Species	John S. Wilkins	9781108987196
Understanding the Nature–Nurture Debate	Eric Turkheimer	9781108958165
Understanding How Science Explains the World	Kevin McCain	9781108995504
Understanding Cancer	Robin Hesketh	9781009005999
Understanding Forensic DNA	Suzanne Bell and John Butler	9781009044011
Understanding Race	Rob DeSalle and Ian Tattersall	9781009055581
Understanding Fertility	Gab Kovacs	9781009054164

Understanding Genes

KOSTAS KAMPOURAKIS
University of Geneva

CAMBRIDGE
UNIVERSITY PRESS

CAMBRIDGE
UNIVERSITY PRESS

University Printing House, Cambridge CB2 8BS, United Kingdom

One Liberty Plaza, 20th Floor, New York, NY 10006, USA

477 Williamstown Road, Port Melbourne, VIC 3207, Australia

314–321, 3rd Floor, Plot 3, Splendor Forum, Jasola District Centre,
New Delhi – 110025, India

103 Penang Road, #05–06/07, Visioncrest Commercial, Singapore 238467

Cambridge University Press is part of the University of Cambridge.

It furthers the University's mission by disseminating knowledge in the pursuit of
education, learning, and research at the highest international levels of excellence.

www.cambridge.org
Information on this title: www.cambridge.org/9781108835473
DOI: 10.1017/9781108884150

First published 2022

Printed in the United Kingdom by TJ Books Limited, Padstow Cornwall

A catalogue record for this publication is available from the British Library.

Library of Congress Cataloging-in-Publication Data
Names: Kampourakis, Kostas, author.
Title: Understanding genes / Kostas Kampourakis, Université de Genève.
Description: Cambridge, UK ; New York, NY : Cambridge University Press, 2021. | Series:
Understanding life | Includes bibliographical references and index.
Identifiers: LCCN 2021007095 (print) | LCCN 2021007096 (ebook) | ISBN 9781108835473
(hardback) | ISBN 9781108812825 (paperback) | ISBN 9781108884150 (ebook)
Subjects: LCSH: Genes – Popular works. | Genomics – Popular works. | Genetics – Popular works.
| BISAC: SCIENCE / Life Sciences / Genetics & Genomics | SCIENCE / Life Sciences / Genetics &
Genomics
Classification: LCC QH447 .K3625 2020 (print) | LCC QH447 (ebook) | DDC 572.8/6–dc23
LC record available at https://lccn.loc.gov/2021007095
LC ebook record available at https://lccn.loc.gov/2021007096

ISBN 978-1-108-83547-3 Hardback
ISBN 978-1-108-81282-5 Paperback

"*Understanding Genes* is an essential guide to this important, complex, and sometimes incendiary topic. In his clear and balanced discussion, Kostas Kampourakis cuts through all the hype and misconception that often surround the debate about what genes are and what they do, and provides the most honest and careful discussion I have seen of how DNA participates in the processes that support life. In doing so, he reveals the real promise, limitations, and dilemmas of the current age of genomics."

Philip Ball, science writer and author of *How to Grow a Human*

"Did you know that two blue-eyed people can have a brown-eyed child? Why calico cats are (almost) always female? It's in the genes, but it's not *all* in the genes. Kampourakis shows that, while genes are unquestionably important, fears of 'designer babies' are both overblown and misguided. Genes alone do not make you who you are. They are not the ultimate essence of life. *Understanding Genes* is simply the best book out there for students or anyone wanting a smart, thoughtful introduction to what genes are and do – and what they aren't and don't."

Nathaniel Comfort, Professor, Department of the History of Medicine, Johns Hopkins University

"Kampourakis has produced a comprehensive but highly readable introduction to genetics and genomics. His take on the fallacy of genetic fatalism is a must-read for both geneticists and the casual reader . . . The role of genetics and genomics in society is treated comprehensively by Kampourakis. He has produced a very readable book with an important message about genetic fatalism – it doesn't exist!"

Professor Robert DeSalle, American Museum of Natural History, New York

" . . . provides a plain, rich, and direct narrative of what a gene is and is not, with practical examples of how genes relate to our daily life . . . clearly identifies controversial views in [the] fields of genetics, genomics, cell and organismic biology, and clarifies them for the comprehension of the just initiated as well as the experienced reader."

Carlos Sonnenschein MD, Tufts University School of Medicine, Boston, MA, USA, and Centre Cavailles, École Normale Supérieure, Paris, France

"*Understanding Genes* is a remarkably clear, rigorous, and yet accessible review of the biological and social roles of genes. Building on a wide range of sources including history, biology, philosophy, and social studies, the book identifies a variety of gene concepts currently in use, illustrates their significance through a wealth of concrete examples, and discusses the relations between these different ways of understanding genes. By deftly combining conceptual analysis with empirical evidence, the book succeeds in comprehensively introducing this complex subject without oversimplifying. It is highly recommended to readers venturing in this domain for the first time, as well as to experts wishing to expand their perspective."

Sabina Leonelli, University of Exeter, UK

"Genes – many people use the word, few understand its many meanings and how they changed over time: from tools to think with, to tools to trace ancestors with. This book guides the reader through the many transformations of this concept from conception to misconceptions, from Mendel to the media. We learn about genetics, genomics, and post-genomics, but also about the interactions between scientific and public understandings and the role of metaphor in spicing things up. Readers come to realize that genes are neither essences, nor things, nor actors; genes only work in context, and in collaboration with each other within an interactive genome. This makes it difficult to find easy solutions to medical problems, but it also means that genes don't determine who we are. This book is more than a guide to understanding genes; it is essential reading for everyone interested in the role that genes play in science and culture."

Brigitte Nerlich, University of Nottingham, UK

"In rigorous but uncomplicated prose, Kostas Kampourakis gives us a present we wish we could have received 100 years ago: a clear explanation of what genes do, what they do not do, what they are, and what they are not. In doing so, he teaches us salutary lessons in both the history and philosophy of science and in human psychology. At a time when our ability to manipulate nature is reaching new levels, Kampourakis provides a road map for understanding the relevance of genetics to our lives. This is a book everyone should read."

Oren Harman, Senior Research Fellow at the Van Leer Jerusalem Institute and Chair of the Graduate Program in Science, Technology and Society, Bar Ilan University, and author of *The Man Who Invented the Chromosome*, *The Price of Altruism*, and *Evolutions: Fifteen Myths that Explain Our World*

"*Understanding Genes* is the first book that provides an honest, nuanced, and full accounting of how genes operate in an organism that is accessible to a general reader. I have not seen in one volume such clear analysis of the 'gene' and its deconstruction from a primary cause to a 'segment of DNA' that is a necessary, but not sufficient, cause of different types of biochemical events. The book exhibits the expertise of an author whose breadth of knowledge of genetics, history and philosophy of science, and science education makes this book exceptionally valuable as a scientific antidote to the tide of popular oversimplifications and the trend in the scientific literature of genetic reductionism."

Sheldon Krimsky, Lenore Stern Professor of Humanities & Social Sciences, and Adjunct Professor of Public Health & Community Medicine, Tufts University

"If you are looking for a concise and up-to-date book on the role of genes (and the science of genes) in our society, look no further: *Understanding Genes* is an accessible, yet nuanced, account of how the concept of the gene has developed throughout history, how its cultural and social meanings have changed, and how genetic factors influence the expression of human behavior and diseases. It conveys not only the basics of genetic thinking, but also a sense for how our understanding of what genes are, and what they do, is always also a response to the big questions that society asks at any given time. I highly recommend this beautifully written book to students, journalists, researchers from other disciplines, and in fact anyone seeking to understand the role of genes – and of genetics – in our world."

Barbara Prainsack, University of Vienna, Austria

"In *Understanding Genes*, Kostas Kampourakis draws on history and popular culture as well as the latest scientific research to help the beginning reader to grasp what genes are, why they are important, and how to give that importance its due without hype or hysteria. Anyone looking for an introduction to genetics that is both reliable and readable need look no further."

Gregory Radick, University of Leeds, UK

"This excellent book is comprehensive, detailed, and amazingly informative, yet eminently readable; it's a really lovely synthesis of the past half-century of thought about what genes are, what genes do, and why they – along with their contexts – are so extremely important. Kampourakis presents biological facts with a 'systems' perspective that remains unwaveringly attentive to the fact that genetic information is always embedded in a context, a context that renders developmental outcomes unpredictable from DNA sequence information alone. Because the book holds fast to this valuable perspective, it brilliantly and clearly makes the case that there are no such things as genes '**for**' specific traits or diseases, despite what we might have gleaned from other media. By deploying wonderful new metaphors and unpacking older and potentially misleading ones, Kampourakis helps readers to avoid many of the misunderstandings that arise from various sources. Accurate and poised at the cutting edge, this primer is lucid enough to be accessible for the general public and students learning about genetics for the first time, but erudite enough for scientists interested in what we currently know about genes. *Understanding Genes* beautifully illustrates the shortcomings of the Human Genome Project, genome-wide association studies, and current personalized medicine and direct-to-consumer genetic tests, clarifying what we now understand and what we are still very much in the dark about. This book serves as an important antidote to the optimistic hype that has a lot of people believing that treatment programs based on individuals' DNA are just around the corner. They are not, and this book explains why, making it a truly *important* read for everyone – or at least, everyone who has genes."

David S. Moore, Pitzer College and Claremont Graduate University

To my brother, Yiannis, and our mother, Evaggelia, who have always made me think hard about "nature" and "nurture"

Contents

Foreword	*page* xiii
Preface: Genes, Science, and Science Fiction	xv
Acknowledgments	xx

1 The Public Image of Genes — 1
Genes in the Media — 1
The Public Perception and Understanding of Genes — 11
The Impact of Media Representations on Understanding:
　　Angelina Jolie's Double Mastectomy — 17
Genetic Disease and Risk Literacy — 23

2 The Origin and Evolution of the Gene Concept — 31
Mendel and the Gene Concept — 31
The "Classical" Gene — 41
The "Molecular" Gene — 48
The "Developmental" Gene — 57

3 The Devolution of the Gene Concept — 65
The "Exploded" Gene — 65
The Human Genome Project Gene — 72
The ENCODE Gene — 79
Genome-Wide Association Studies and the "Associated" Gene — 86

4 There Are No "Genes For" Characteristics or Disease — 96
Biological Characteristics: Eye Color and Height — 96
Behavioral Characteristics: Aggression — 101

Monogenic Diseases: Thalassemias and Familial
  Hypercholesterolemia   105
Multifactorial Diseases: Cancers   113

5 **What Genes "Do"**   **121**
The Development of Characteristics in Individuals   121
Genes are Implicated in the Development of Characteristics in
  Individuals   130
The Variation of Characteristics in Populations   138
Genes Account for Variation in Characteristics in Populations   145

6 **The Dethronement of Genes**   **149**
Genes Are Not Master Molecules: The Relation Between
  Genes and Characteristics or Disease Is Complex   149
Genomes Are More than the Sum of Genes: Epigenetics   153
Association Does Not Equal Causation: Genetic Tests and the
  "Associated" Gene   162
Genes Are Not "Texts" Waiting to be Read: Beware
  of Gene Metaphors   172

Concluding Remarks: How to Think and Talk about Genes?   **182**

Summary of Common Misunderstandings   187
References   189
Index   207

Foreword

Following on from the success of his *Understanding Evolution* book, Kostas Kampourakis and I discussed the need for a book that addressed the common misconceptions surrounding genes in a comparable way. *Making Sense of Genes* resulted and was published in 2017, receiving universally positive feedback for its eloquent, multidisciplinary treatment of the fundamental questions: What are genes and what do they do? From the success of this book came an invitation for Kostas to speak at the Cambridge Science Festival; it was an honour to host him in Cambridge and, as we walked to the venue, share with him his book on display in the window of the Cambridge University Press bookshop.

Since then Kostas and I have developed the *Understanding Life* series together. Our vision for it is to provide concise, accessible guides to key topics, written by leading thinkers in the field and focusing on the common misconceptions and misunderstandings that are potential barriers to gaining a deeper understanding. Genes are, of course, an obvious topic for inclusion in this series, and hence this book arose, which updates and condenses the coverage in *Making Sense of Genes*, providing an invaluable introduction to the topic. Its discussion of the role of genetics and genomics in society, and their presentation in the media, are particularly timely.

It has been an enormously fulfilling, fun, and enjoyable experience working with Kostas on this book and on the series more broadly. I'm proud of what we have achieved with it and it makes an important contribution to the Press' mission, to publish and disseminate high-quality information to aid learning. I very much look forward to seeing how *Understanding Life* develops and grows in the future as I move on to new adventures.

Dr. Katrina Halliday
Executive Publisher, Life Sciences
Cambridge University Press

Preface: Genes, Science, and Science Fiction

Antonio and Marie Freeman follow a nurse who shows them the way to the geneticist's office. Marie carries their young son, Vincent. As soon as they sit down, a monitor turns on and they see four embryos, each consisting of a few cells. Next to another monitor across the room, the geneticist is sitting. He says: "Your extracted eggs, eh ... Marie, have been fertilized with Antonio's sperm. After screening we are left, as you see, with two healthy boys and two very healthy girls. Naturally, no critical predispositions to any of the major heritable diseases." The geneticist stands up and approaches the couple. "All that remains is to select the most compatible candidate," he says as he sits down next to them. "First, we might as well decide on gender – have you given it any thought?"

"We would want Vincent to have a brother, you know, to play with," Marie says, looking at her son, who is playing on the ground with a ball-and-stick molecular model.

"Of course you would. Hello Vincent," the geneticist says, smiling at him. Vincent smiles back and shyly says "Hi."

The geneticist turns to the couple and continues: "You have specified hazel eyes, dark hair, and fair skin. I have taken the liberty of eradicating any potentially prejudicial conditions: premature baldness, myopia, alcoholism and addictive susceptibility, propensity for violence, obesity, etc."

"We didn't want any, I mean diseases yes, but ... " Marie interrupts the geneticist, and she and Antonio look at each other. Antonio continues:

"Right, we were just wondering if it's good to just leave a few things to chance."

The geneticist replies immediately: "You want to give your child the best possible start. Believe me, we have enough imperfection built in already. Your child doesn't need any additional burdens. Keep in mind, this child is still you. Simply the best of you. You could conceive naturally a thousand times and never get such a result."

This is a scene from the 1997 dystopian film *GATTACA*, which presents an oppressive society where sexual intercourse between a man and a woman is no longer the natural way of having a child. Instead, *in vitro* fertilization and genetic screening of the emerging embryos are invoked to ensure that only children with the desired characteristics and without any "prejudicial conditions" are brought to life. Vincent, who was not conceived through such a process, is considered to be genetically inferior. This is why Marie and Antonio decided to have a second child through *in vitro* fertilization and genetic screening.

GATTACA was a science fiction film that postulated a future society of genetic discrimination. At that time, the Human Genome Project was still under way and the potential for genetic screening was relatively limited – even though the expectations were great. Preimplantation genetic diagnosis was possible at the time, but it was usually used for particular genetic diseases in the cases of couples with a family history of those diseases (diagnosis is the search for conditions or specific alleles – different versions of the same gene – in people already considered as likely to have them, whereas screening refers to the search for conditions or specific alleles in the general, healthy, and asymptomatic population).

Nowadays, almost a quarter-century later, things seem to have changed. According to one estimate by Antonio Regalado, senior editor for biomedicine for the *MIT Technology Review*, more than 26 million people had taken a genetic test by the beginning of 2019. Margo Georgiadis, president and chief executive officer of Ancestry.com, has estimated that by early 2020, 30 million people had taken a DNA test. Even though in most of these cases the tests were related to ancestry, one could expect prospective parents to be interested in predictive genetic tests to select the healthiest, smartest, loveliest – or whatever – babies. Does this mean that we are entering a *GATTACA*-style era? How far are we from such possibilities? This might happen if indeed genes somehow cause conditions in an if-you-have-the-gene-then-you-will-have-the-condition manner. In that case, we could indeed screen embryos and select only those with the desired characteristics. If we came to know the "genes for" particular characteristics, then we could distinguish between individuals who carry them and those who do not. Whether this is or will become possible (or even plausible) requires understanding genes: what they are and are not, as well as what they can and cannot "do."

The aim of the present book is to counter misperceptions about genes, and help readers to acquire a better understanding of them. Along the way I show that the importance of genes has often been exaggerated. Of course, I am not going to argue that genes are not important – indeed, they are! But it is one thing to say that genes are important for what we are or do, and another that they matter more than anything else. Genes have been presented as autonomous entities that contain all the necessary information to determine characteristics and are capable of making use of it. They have thus been described as the "essence" of life, as the absolute "determinants" of characteristics and disease, and therefore as providing the ultimate explanations for all biological phenomena because the latter can be "reduced" to the gene level and thus be explained.

Therefore, there exist at least three misunderstandings about genes:

Genetic essentialism: Genes are fixed entities that are transferred unchanged across generations, and that are the essence of what we are by specifying characteristics from which their existence can be inferred.

Genetic determinism: Genes invariably determine characteristics, so that the outcomes are just a little, or not at all, affected by changes in the environment, or by the different environments in which individuals live.

Genetic reductionism: Genes provide the ultimate explanation for characteristics, and so the best approach to explain these is by studying phenomena at the level of genes.

These definitions help us distinguish between three important properties usually attributed to genes: (1) that they are fixed essences that specify who we are; (2) that they alone determine characteristics notwithstanding the environment; and (3) that they best explain the presence of characteristics. Collectively, these ideas can be described as *genetic fatalism*. In the present book, I explain why genetic fatalism is wrong.

A central feature of the present book is that it is mostly about human characteristics and disease. When this is not the case, it is usually about phenomena of relevance to human life. I must note that this is not due to any anthropocentricism on my part. Quite the contrary, I believe that we are not anything special in this world, or at least that we are not any more special than any other organism that lives in it. Nevertheless, I thought that the book would be more interesting and comprehensible to readers if I discussed phenomena about, or relevant to, human life. This approach is biased, of course, because it overlooks important aspects of life on earth. I hope that readers will find this biased-toward-humans book interesting and didactic. But they should also keep its bias in mind and avoid unwarranted generalizations from the mostly medical-centered and human-focused research presented in this book.

Some basic nomenclature: Gene and protein symbols in humans are written with uppercase letters. However, gene symbols are written with italicized characters, whereas protein symbols are written with regular characters (the protein called "hemoglobin A" is written HBA, whereas the respective gene is written *HBA*). For the purpose of consistency, most gene symbols and gene names in this book are derived from the Human Genome Organization Gene Nomenclature Committee website (www.genenames.org).

A note about historical periods: There are several ways to distinguish between the various periods of research related to genes and genomes. In many cases, and for many reasons, some periods may actually overlap for a significant amount of time. However, simply for convenience, in the present book I consider four distinct historical periods:

1. the period until the coining of the gene concept in 1909, which I describe as the pre-genetics era (even though the term "genetics" was coined in 1906);
2. the period from 1909 until the initiation of the Human Genome Project in 1990, which I describe as genetics because the focus of research was

mostly on genes – even though the Human Genome Project itself initially focused on individual genes. This period can be roughly divided into two periods, classical genetics (1900s to 1950s) and molecular genetics (1950s to 1990s);

3. the period from 1990 to 2013, when the findings of the ENCODE project and of genome-wide association studies made clear the vast complexity of genomes, which I describe as genomics; and

4. the period from 2013 to today, which I describe as postgenomics.

I must note that I do not claim that this is how the history of genetics and genomics should be perceived or organized. I only want to note that there has been a shift of focus from genes to genomes, as well as that it took us a whole century – the century of the gene, as historian Evelyn Fox Keller described it – to realize the complexities of heredity.

The present book is intended for anyone who wants an accessible but rigorous introduction to genes. It provides a concise overview of contemporary genomics research and concepts. This research advances at an extremely fast pace, and I am sure that now that you are reading this book, there exist new articles and books that I could have considered. This was already the case when I was working on this new 2021 edition of the book, less than four years after the publication of the original 2017 edition. Nevertheless, I am confident that the main points of the present book and its conceptual foundations will remain unchanged for many years to come. Let us now begin our quest to understand genes.

Acknowledgments

There are many people I would like to thank because they made writing this book possible in various ways. But there is no-one else that deserves to be acknowledged more in this case than Katrina Halliday, executive publisher for the life sciences at Cambridge University Press. Neither this book as you see it nor the book series to which it belongs would have existed without the insight and support of Katrina. The first edition of the present book, published in 2017, was very well received and was commended for its quality and readability (see excerpts from and links to the reviews at http://kampourakis .com/making-sense-of-genes). Yet, that was still an academic book. Thanks to Katrina, we now have this revised and updated, but also concise, version that I hope you will appreciate.

I am grateful to Bruno J. Strasser and Andreas Müller, who support my work and research at the University of Geneva. My interest in human genetics goes back in time to when, as an MSc student, I had the opportunity to work at the laboratory of Emmanouil Kanavakis at the University of Athens, whom I thank for that opportunity. While writing this book, I have been very fortunate to benefit from the thoughtful feedback of several scholars: Garland Allen, John Avise, Sheldon Krimsky, Alessandro Minelli, David Moore, Staffan Müller-Wille, John Parrington, Giorgos Patrinos, Erik Peterson, Anya Plutynski, Gregory Radick, Andrew Reynolds, Carlos Sonnenschein, Eric Turkheimer, and Tobias Uller. I thank them all for their valuable comments and suggestions. I owe special thanks to Nathaniel Comfort, whose comments were extremely useful in clarifying the main argument of this book. Writing this book has also benefited from discussions during an older collaboration with

Richard Burian. I am also grateful to Olivia Boult and Sam Fearnley at Cambridge University Press for their work toward the publication of this book, as well as to Gary Smith for his meticulous copy-editing. Finally, I thank Nikos Moschonas and Marilena Papaioannou, who notified me about some minor issues that they identified while working on the Greek translation of the 2017 edition. Of course, as is always the case, I am responsible for any remaining problems or errors.

Last but not least, there are always those to whom I owe a lot: my family. I dedicate this book to my brother and our mother because the striking differences and similarities among us have always made me think hard about "nature" and "nurture."

1 The Public Image of Genes

Genes in the Media

This chapter is about the public image of genes. But what exactly do we mean by "public"? Here, I use the word as a noun or an adjective vaguely, in order to refer to all ordinary people who are not experts in genetics. I thus contrast them with scientists who are experts in genetics – that is, who have mastered genetics-related knowledge and skills, who practice these as their main occupation, and who have valid genetics-related credentials, confirmed experience, and affirmation by their peers. I must note that both "experts" and "the public" are complex categories that depend on the context and that change over time. There is no single group of nonexperts that we can define as "the" public, as people around the world differ in their perceptions of science, depending on their cultural contexts. We had therefore better refer to "publics." The differences among experts nowadays might be less significant than those among nonexperts, given today's global scientific communities, but they do exist. Finally, both the categories of experts and publics have changed across time, depending, on the one hand, on the level of experts' knowledge and understanding of the natural world, and, on the other hand, on publics' attitudes toward that knowledge and understanding.

This bring us to another important question: What is the relation between experts and publics? A long-held view is the so-called deficit model. According to this, scientific knowledge and understanding are transmitted by the enlightened experts to the ignorant publics, in an attempt by the former to educate the latter. This is a view in which experts always have superior status compared to publics. However, this is far from accurate. Both the way science itself is conducted and the way its findings are communicated have never been completely separated from their social contexts. In general, one might argue that science and society do not simply interact, but are co-constructed; science is done within society and cannot be demarcated from it. Therefore, the communication of the conclusions of scientists to the various publics is not a linear process of transmission. Rather, it is a process of constant interaction and negotiation.

In her detailed account of the popular images of genetics throughout the twentieth century, media scholar José van Dijck has shown that there has never been a clear separation between science and its images, in the same sense that there have never been clearly separated scientific and commercial or public and private domains. Thus, she argued that the mediation of science has not been the outcome of interactions between demarcated communities: the scientists who command knowledge and the journalists who command its public representation. Rather, the mediation of science has been the outcome of interactions "between various professional groups, who are not merely facilitators or manipulators of expert knowledge, but who are themselves active participants in a public definition of science." Images of science are never mere illustrations of scientists' practices, nor are imaginations mere reflections of people's anxieties about these practices. Rather, images and imaginations are rhetorical tools in the construction of a public meaning, which are intricately connected. Van Dijck described the outcome of this connection as "imagenation," noting that "Rather than a linear diffusion of knowledge, 'imagenation' assumes a recursive circular transformation of knowledge." This circularity describes the "multi-layered dissemination of genetic knowledge." As media and film scholar Kate O'Riordan has nicely put it, "It might be helpful to take media audiences as publics orientated towards mediated technoscience, rather than seeing audiences as orientated towards the technoscience of genomics through media." In short, knowledge about genes and genomes is not simply diffused from expert-producers toward the nonexpert-consumers through the media. Rather, the media actively participate in the public representation of this knowledge. With this in mind, let us now look at how genes have been represented in the media.

If you look at media headlines, you will find several accounts of how genes affect various aspects of our lives. The general message conveyed in many cases is that there exist "genes for" characteristics. That genes affect biological characteristics – such as the color of our hair, eyes, or skin – is not news, of course. What is news, and what often features in headlines, is that genes also affect behaviors or life outcomes. For instance, an article on the CNN website titled "The star gene: next generation celebrity" includes photos of famous parents and children such as Kirk Douglas and his son Michael Douglas, Judy Garland and her daughter Liza Minelli, Henry Fonda and his children Peter and Jane Fonda, Martin Sheen and his sons Emilio Estevez and Charlie Sheen, Jon Voight and his daughter Angelina Jolie, and many, many more. What might the title of this article imply? That there exists a "gene for" becoming a Hollywood star. Aren't you tempted to think that, besides the morphological similarity that is evident in many of these parent–child cases, there is also something else, like acting talent, that runs in families? As the CNN article states, there is: the "star gene."

Other news articles make similar claims, reporting conclusions from research in genetics. For instance, an article in the *Financial Times* titled "Genes determine how young use internet and social media" reported that "Genes play an unexpectedly big role in determining how young people use the internet and social media, according to a large UK study of 16-year-olds." Reporting on the same UK study, another article in *Science Daily*, under the title "Online media use shows strong genetic influence," suggested that "Online media use such as social networking and gaming could be strongly influenced by our genes." Genes have also been reported to impact financial success. This was suggested by an article in the *Daily Mail* titled "Being rich and successful really IS in your DNA: Being dealt the right genes determines whether you get on in life," and by an article in *The Times* titled "Scientists find 24 'golden' genes that help you get rich" (these two articles reported on different scientific studies). Could there be a "gene for" using social media or being rich?

And there is more. Did you know that your romantic life also seems to be affected by your genes? If you have a happy marriage, it may be due to your genes. "Key to a happy marriage? It's in your genes, scientists discover," an article in the *Telegraph* informs us. "This gene could be the secret to a happy marriage: study," we read in the *New York Post*. Both of these articles reported on a study suggesting that people with a specific genotype (that is, a particular combination of alleles) were more likely to report higher satisfaction in their marriages. But what if your marriage is not a happy one? Again, genes may have the answer, because "Infidelity lurks in your genes," according to the *New York Times*. This article reported on a study that found that "Women are more likely to cheat on their partner if they carry the 'infidelity gene'," as the *Daily Mail* also reported. And if you have no relationship at all, no worries! Companies like Gene Partner can analyze your DNA and find the perfect match for you because, as they state on their webpage, "Love is no coincidence!" What they do is "Matching people by analyzing their DNA."

What is the message conveyed by media articles like these? Whether you have a happy, romantic relationship, an unhappy one, or no relationship at all may not be due to your choices or to those of your (actual or potential) partners. Whether you are rich or not may not be due to the hard work you did or did not do, or the circumstances you happened to experience or not experience. Whether your adolescent child spends a lot or limited time on social media may not be due to your parenting or to what they see their friends and other people doing. Whatever you did or did not do, whatever you could or could not do, may not be that important; genes are presented as the main causal factors for any of these life outcomes. The attribution of such outcomes to genes is actually a win–win situation. On the one hand, you are not to blame if you do not have a happy marriage, if you did not become rich, or if your child is addicted to social media, because there was nothing you could do – it is in the genes. On the other hand, other people or society at large are not to blame for how they treated you, for the opportunities they did not give you, or for the prevalent models that influenced your child, because there was nothing you could do – again, it is in the genes.

Several commentators have long argued that such media representations of genes can be misleading, and can perpetuate inaccurate conceptions about what genes are and, especially, what they can do. In 1991, epidemiologist Abby Lippman coined the term "geneticization" to describe the phenomenon of making overt attributions to genes:

> Geneticization refers to an ongoing process by which differences between individuals are reduced to their DNA codes, with most disorders, behaviours and psychological variations defined, at least in part, as genetic in origin … Through this process, human biology is incorrectly equated with human genetics, implying that the latter acts alone to make us each the organism she or he is.

Quoting Lippman, biologist Ruth Hubbard presented a book-length account of several facets of our lives in which geneticization seems to prevail: our characteristics, disease, behaviors, education, employment, and more. In the afterword of that book she noted that

> of course, everything that happens in our lives has a "genetic component." But so what? The fact that everything we are and do involves genes in no way implies that knowing everything about their location, composition and the way they function will enable us to understand all of human health, and to predict, prevent, or control all diseases, and unwanted behaviors.

But why has so much power been attributed to genes?

Sociologists Dorothy Nelkin and Susan Lindee, in their analysis of the public representations of the gene, argued that "the gene of popular culture is not a biological entity. Though it *refers* to a biological construct and derives its cultural power from science, its symbolic meaning is independent from biological definitions. The gene is, rather, a symbol, a metaphor, a convenient way to define personhood, identity, and relationships in socially meaningful ways." According to Nelkin and Lindee, the images and narratives of the gene in popular culture convey a message that they call genetic essentialism, which "reduces the self to a molecular entity, equating human beings, in all their social, historical, and moral complexity, with their genes." According to them, "Today these narratives [of mass culture] present the gene as robust and the environment as irrelevant; they devalue emotional bonds and elevate genetic ties; they promote biological solutions and debunk social interventions." This simply means that to a large extent who we are and what we do are largely determined by our genes; non-genetic influences such as environmental ones do not matter. This is a view of genetic determinism (though not explicitly defined as such) that is according to Nelkin and Lindee predominant in popular culture.

However, not all agree with the view that genetic determinism messages are widespread in popular culture. Rhetorical criticism scholar Celeste Condit, and her colleagues, conducted a systematic analysis of 653 magazine articles published during the twentieth century in the USA, in order to assess whether or not they conveyed messages about genetic determinism, defined as "the assignment of exclusive influence over human outcomes to genes." They divided the twentieth century into four periods, having found that different metaphors about genes predominated in each of these periods. According to their analysis, the messages conveyed in magazines have not been more deterministic in more recent times than in the past. Nor has determinism ever been the most prevalent message, as there have been more statements about an influence by both genes and environment than about the influence of genes alone in all periods (Table 1.1). Therefore, genetic determinism was not the prevalent message in magazine articles in the twentieth century, at least in the USA, according to this study.

Another study analyzed how the gene concept has been presented in major national newspapers from the USA, the UK, France, and Norway. Science communicator Rebecca Bruu Carver and her colleagues analyzed how the gene was represented in 600 randomly selected, gene-related articles published between July 2005 and July 2008. The framework they used distinguished between the five following ways of framing the gene concept: (1) symbolic, referring to an abstract or metaphorical representation of inheritance; (2) deterministic, referring to a definite causal agent that might even act against environmental factors; (3) relativistic, referring to a predisposing factor; (4) materialistic, referring to a discrete physical unit; (5) evolutionary, referring to the central object of evolution, a marker for evolutionary change, or a factor that interacts with the environment. Carver and colleagues found that there was no overrepresentation of the deterministic frame as it was found in only one-sixth of the articles (Table 1.2). The authors concluded that older accounts of genetic determinism largely concerned symbolic representations of the gene concept, whereas actual claims of genetic determinism were not common in public discourse. In other words, recent newspaper

Articles with statements	1919–1934		1940–1954		1960–1976		1980–1995	
	n	*%*	*n*	*%*	*n*	*%*	*n*	*%*
No influence by genes (or pro-environment)	7	5	1	1	0	0	0	0
Influence by both genes and environment	79	54	44	61	60	61	60	66
Influence by genes only (or against environment)	45	34	27	38	39	40	31	40

Columns do not add to the total number shown or to 100 percent because some articles may include more than one or none of the statements.

Source: Adapted from *The Meanings of the Gene: Public Debates about Human Heredity* by C. M. Condit. Reprinted by permission of the University of Wisconsin Press. © 1999 by the Board of Regents of the University of Wisconsin System. All rights reserved.

Table 1.1   Genetic determinist statements in magazine articles published in the USA during the twentieth century

accounts in the USA, the UK, France, and Norway of what genes are and what they do have not been overtly deterministic.

Other science communication scholars have explored the representations of genetics in popular culture, especially in novels and films. Science communication scholar David Kirby has analyzed several science fiction films produced during the twentieth century, focusing on their treatment of eugenics. According to his analysis, during the first period (1900–1929) many films uncritically accepted "the eugenicist's conception of humanity's tainted animal heritage," while at the same time warning that any attempt to alter human nature is either doomed to fail or to create soulless monsters such as those in *Frankenstein* and *Dr. Jekyll and Mr. Hyde*. The second period (1930–1949) is characterized by the same ideas, with films having two

Frame	Symbolic	Deterministic	Relativistic	Materialistic	Evolutionary
Percentage	31.8	16.2	13.5	25.6	12.9
Example	"I have inherited the shopping gene from my Mom."	"Researchers have found the gene for breast cancer."	"Genes increase risk of developing cancer."	"NPC is caused by a mutation in a gene on chromosome 18. Children with the disease have inherited two copies of the abnormal gene."	"Comparison between human and ape DNA reveals that some human and ape genes evolved very swiftly."

Source: Carver, R. B., Rødland, E. A., and Breivik, J. (2013). Quantitative frame analysis of how the gene concept is presented in tabloid and elite newspapers. *Science Communication*, 35(4): 449–475.

Table 1.2   Gene frames related to article topic (600 articles)

main themes: initially "mad evolutionist" characters who design evil experiments to show humanity's connection to the animal world, and later Nazi-like mad scientists who aim to create super soldiers. The films of the third period (1950–1969) are characterized by concerns about a nuclear war and the subsequent effects of radiation, with very few films making any reference to DNA. During the fourth period (1970–1989), films focus extensively on genetic engineering and recombinant DNA technology, with the latter representing the most important threat. Finally, films during the last period (1990–2004) suggest that identity resides in genes and that any attempts to alter the genome would fundamentally change it. Despite the differences in the main themes and messages of the films of these periods, Kirby concluded that they almost uniformly convey the message that our fundamental nature lies within our genome, with the implication that this nature could be improved by genetic engineering. However, these very same films are critical toward any such kind of intervention by technological means. In this sense, the message conveyed is that the genome is sacred and so we should refrain from making any changes to it because we would thus alter its authenticity.

Film and literature scholar Everett Hamner has provided a detailed analysis of science fiction novels and films from the 1960s to very recently, identifying three kinds of narratives: (1) genetic fantasy, in which a new finding or tool is considered in distant-future or super-hero stories with the aim of commenting on the current situation; (2) genetic realism, where science fiction inspires technically detailed and plausible scenarios; and (3) genetic meta-fiction, where the fantastic and real are blurred. According to Hamner, genetic fantasy emerged first during the 1960s and the 1970s, when speculation about genetic recombination was popular. Genetic realism grew out of genetic fantasy after the Human Genome Project, during the 1990s and the 2000s, when technologies related to genes had advanced. Finally, genetic meta-fiction emerged more recently from the other two genres, describing a self-awareness about gene testing and editing. Hamner noted that these three narratives did not replace one another but rather overlapped, resulting in a cumulative rather than a successive trend. Differences notwithstanding, Hamner showed that in novels from all three genres there is a tendency to resist genetic determinism, for instance, by highlighting human uniqueness even in the case of clones, by rejecting the notion that genes determine fate, or by also considering the role of environment and culture as well as chance and choice. Overall, the idea of genetic determinism exists in science fiction novels and films, but it is often questioned.

How about television? Film scholar Sofia Bull has analyzed the representation of genetics in TV series and shows such as *CSI* and *House*, as well as various documentaries, sitcoms, and genealogy reality shows in the USA and the UK. Her main conclusion has been that in the beginning of the twenty-first century, notions of uncertainty and complexity, and ideas about the modifiability of biological processes and bodies, have gradually come to coexist with the older, established essentialist, determinist, and reductionist notions about DNA. As Bull argued, and showed with various examples, television functions as a cultural forum on genetics that stages multifaceted negotiations between long-standing essentialist ideas and the new genetics. For instance, genealogy TV shows convey the message that kinship and ancestry are ultimately located in, and determined by, genes. Bonds between "blood relatives" are overemphasized as they are considered to be more *real* or *true* than other social affiliations. This is based on an essentialist and determinist understanding of the genome as containing the blueprint of both identity and relatedness. However, on several occasions, programs also present insights from research in epigenetics (see Chapter 6) that highlight the complex and dynamic nature of genetic ancestry. Bull concluded that "Although essentialist perspectives have remained prominent on television, particularly across forensic crime procedurals, genealogy TV and family-centric reality shows ... distinctive elements of television's visual form, narrative structure, production, distribution and reception has made it a key site for gradually imagining a more complex and indeterminate (molecular) world."

Overall, one can conclude that whereas ideas about genetic determinism and genetic essentialism (and sometimes genetic reductionism) exist in the media, they are not the predominant ones and tend to coexist with ideas about complexity and multicausality, especially in the era of genomics. There exist, of course, individual cases where the impact of genes is exaggerated, as in the examples that I presented in the beginning of this section, but these are not the only ones. Let us now see what people's beliefs about genes can be.

The Public Perception and Understanding of Genes

It is far from simple and straightforward to describe the public perception and understanding of genes, because most of the available data come from studies in the USA and Europe. As some researchers have nicely put it, most people in the world are not WEIRD: Western, Educated, Industrialized, Rich, and Democratic. According to an estimate, these people represent a minority of the people on earth (about 12 percent of the global human population), but most research has been conducted in these populations. As not many studies reporting data from a variety of countries and cultural contexts exist, conclusions about the public understanding of DNA and genes are not easy to make.

However, there is one recent study that reports the results from a very large public survey on attitudes toward genomic data sharing, which involved 36,268 individuals across 22 countries. Even though this study focused on participants' willingness to donate DNA and medical information, there were some interesting findings at the conceptual level. The researchers asked participants the following question: "Are you familiar with DNA, genetics, or genomics?" Those participants who answered no were classified as being "unfamiliar." Those participants who answered yes were subsequently asked to choose among various reasons for being familiar. Their choices helped researchers distinguish those participants who had personal experience of genetics (being either patients with a genetic disease, or having a family history of a genetic disease, or being professionals who work on genetic disease), who they classified as having "familiarity through personal experience." All the other participants who answered yes to the initial question were classified as having "conceptual familiarity." Overall, the majority of participants across the world stated that they were not familiar with the concepts of DNA, genetics, and genomics: 64.2 percent (23,273 participants) were unfamiliar with these concepts; 25.3 percent (9,182 participants) were familiar in general; and 10.5 percent (3,803 participants) had familiarity through personal experience. Participants' responses for each country are presented in Figure 1.1. With the exception of Italy and the USA, more than half of participants in all other countries stated that they were unfamiliar with DNA, genetics, or genomics. Given that familiarity does not necessarily entail knowledge and understanding – as someone may have heard a term but not know much about it or simply misunderstand it – it is probably the case that fewer people actually knew what DNA, genetics, or genomics are than declared themselves to be familiar with these terms.

The study also investigated participants' perceptions of genetic exceptionalism: whether DNA data are different than other kinds of health data. According to this view, genetic information is special "because it is uniquely identifying, directly links us to our relatives or can provide information about our past, present and future health," and therefore is different than any other kind of medical information. Participants were asked the question "Is DNA information different to medical information – what do you think?" and could choose among three options: "Different," "The same," "I'm not sure." Overall, 53 percent (17,044) of participants viewed DNA as being different from other kinds of medical information. However, there were very different views across the various countries, as shown in Figure 1.1. In Belgium, Canada, France, Germany, India, Japan, Russia, and Switzerland most people thought that DNA information was the same as other kinds of medical information. In all other countries, most people thought it was not, with the exception of the UK, where opinions were exactly divided. Figure 1.1 provides an overview of these data. Several inferences could be drawn, but more detailed research is necessary.

Let us consider another study that looked at the genetics knowledge of participants in the Coriell Personalized Medicine Collaborative. This study involved 4,062 participants in the USA who completed a genetics knowledge and genetics education questionnaire. Among these, 674 were working in the domain of healthcare (HC group) and the other 3,388 had nothing to do with it (NHC group). The findings revealed incoherent understandings. On the one hand, 95.2 percent of the NHC group and 99.6 percent of the HC group answered correctly that the statement "The onset of certain diseases is due to genes, environment and lifestyle" is true. On the other hand, only 66.5 percent of the NHC group and 84.9 percent of the HC group correctly answered that the statement "A 'complex disease' is a health condition brought on by many genes and lifestyle and environment" is true. These differences can be interpreted in various ways. However, if participants' responses are not consistent, then their understanding may not be solid – especially if one considers that only 46.7 percent of the NHC group and 58.8 percent of the

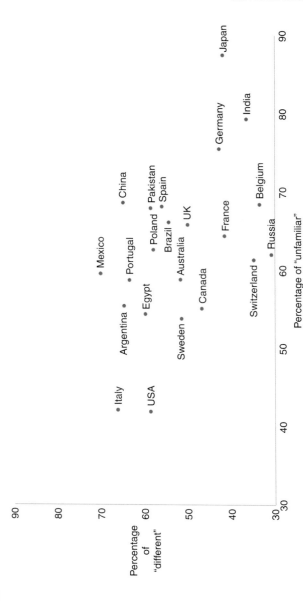

Figure 1.1 Unfamiliarity with DNA, genetics, and genomics, and perception of DNA as being different from other kinds of medical information ("genetic exceptionalism") across 22 countries. Data from Middleton et al. (2020) Global public perceptions of genomic data sharing: what shapes the willingness to donate DNA and health data? *American Journal of Human Genetics*, https://doi.org/10.1016/j.ajhg.2020.08.023.

HC group correctly considered the statement "All body parts have all of the same genes" as true.

Another study in the USA used an online survey to investigate the beliefs of 1,041 participants about genetic and environmental contributions to differences among people for 21 human characteristics (eye color, blood group, color blindness, height, bipolar disorder, schizophrenia, attention deficit/hyperactivity disorder, breast cancer, diabetes, alcoholism, violent behavior, political beliefs, obesity, personality, blood pressure, athleticism, heart disease, musical talent, sexual orientation, intelligence, and depression). Participants were asked to evaluate the basis of these characteristics on a 1–5 scale (only environmental factors = 1; mainly environmental factors = 2; genetic and environmental factors contribute roughly the same = 3; mainly genetic factors = 4; only genetic factors = 5). Participants considered differences in height, eye color, blood group, and color blindness as the characteristics most strongly influenced by genetic factors, with mean scores above 4. Political beliefs were estimated to have the smallest genetic contribution, and was the only one with a mean below 2 (Table 1.3). It is obvious that, except for the biological characteristics and some diseases, participants considered the differences among people for the majority of characteristics as influenced by both genetic and environmental factors. These findings may seem to support the conclusion that beliefs in genetic determinism may not be that high and that people overall believe that both genetic and environmental factors contribute to our characteristics. I should note, though, that acknowledging that multiple factors affect the development of characteristics is not inconsistent with prioritizing genetic explanations. This means that even if people state that both genes and environment influence characteristics, they may still believe that the genes are more important than any other factor.

A different study aimed at exploring how the public interprets news about behavioral genetics, which usually attracts a lot of attention in the news. Overall, 1,413 participants from the USA were given one of three published news articles: 471 were given an article titled "'Liberal gene' discovered by scientists"; 475 were given an article titled "Born into debt: gene linked to credit-card balances"; and 467 were given an article on cancer genetics, titled "Key breast cancer gene discovered" (the latter was the control group of the study). The results supported the conclusion that the readers not only

Characteristic	Mean score ($N = 1,041$)
Eye color	4.65
Blood group (ABO)	4.63
Color blindness	4.44
Height	4.16
Bipolar disorder	3.59
Schizophrenia	3.57
Attention deficit/hyperactivity disorder (ADHD)	3.48
Sexual orientation	3.39
Intelligence	3.34
Breast cancer	3.31
Athleticism	3.20
Heart disease	3.15
Blood pressure	3.03
Diabetes	2.99
Depression	2.94
Musical talent	2.91
Personality	2.74
Alcoholism	2.71
Obesity	2.67
Violent behavior	2.51
Political beliefs	1.70

Only environmental factors = 1; mainly environmental factors = 2; genetic and environmental factors contribute roughly the same = 3; mainly genetic factors = 4; only genetic factors = 5.

Source: Willoughby, E. A., Love, A. C., McGue, M., et al. (2019). Free will, determinism, and intuitive judgments about the heritability of behavior. *Behavior Genetics*, 49(2): 136–153.

Table 1.3 The beliefs of 1,041 US participants about the genetic and environmental contributions to differences among people for 21 human characteristics on a 1–5 scale

generally accepted the information presented to them in the "liberal gene" and "card gene" articles, but also that they inadvertently generalized the influence of genetics to other behaviors that were not mentioned in the articles they read. For example, the participants who read the "liberal gene" article were more likely than the control group to suggest a genetic attribution not only for being liberal or conservative, but also for sexual orientation, mathematical ability, alcoholism, behaving violently, gambling addiction, and having credit card debt. Similarly, the participants who read the "credit card gene" article were more likely compared to the control group to suggest a genetic attribution not only for having credit card debt, but also for all the other orientations, skills, and preferences mentioned. It should be noted that there were no significant differences in the genetic attributions for biological characteristics between the treatment groups and the control group. Overall, one can conclude from this study that people may interpret research on behavioral genetics in an incautious manner, and form beliefs that are not supported by the scientific evidence presented in the articles they had read.

But what exactly is so special about genetic information? Several small-scale studies led by social psychologist Steven Heine have shown that many people hold the view described earlier as genetic essentialism. Heine and colleagues consider essentialism as the idea that all characteristics and properties stem from particular factors, which are internal, existing at the deepest possible levels, and which can be transferred. They have concluded that laypeople apply particular essentialism biases when thinking about genes, which in turn they perceive as being our essences. In this view, genes are:

- immutable – that is, less changeable and more predetermined;
- the ultimate causes, making consideration of other causal factors unnecessary;
- the basis for drawing the boundaries of categories, as members of the same category should have a similar essence and clearly different from the essence of members of another category; and
- natural, in the sense that when a characteristic is attributed to genes, this is how it ought to be.

A first point to note is that the framework of Heine and his colleagues does not distinguish between essentialism, determinism, and reductionism, the way I did in the preface of the present book, and rather lumps them together under the label of essentialism. Beyond this, there is evidence from various small-scale studies that people tend to think in this way. For instance, three studies were conducted in Canada to examine what consequences people draw from a perceived genetic etiology for obesity. In the first study, 131 undergraduates indicated whether or not they believed that obese people can control their weight, as well as that obesity originates from a genetic predisposition or environmental causes. The researchers found an association between a belief in genetic etiology for obesity and a belief that obese people cannot control their weight. These associations were further explored in a second study, in which 143 undergraduates were asked to express their beliefs about an obesity-related phenomenon (metabolic rate) in the light of particular explanations. The researchers concluded that a genetic attribution for high metabolic rate was interpreted as more important than an experiential attribution. Finally, in a third study, 162 undergraduates read one of three fictional media reports presenting a genetic explanation, a psychosocial explanation, and no explanation for obesity. The researchers found that participants who had read the genetic explanation ate significantly more than the others on a follow-up task. These studies support the conclusion that genetic arguments for obesity make people report and act as if genes make obesity a condition beyond one's personal control.

Overall, even if people do not generally prioritize genetic factors over environmental ones, there is evidence that when they are told that genetics is the most important influence, this is a message that they can readily accept. One might argue that genes and DNA are embedded in popular culture in ways that environmental factors are not, as I showed in the previous section. As the studies of media and laypeople's understandings presented so far are independent of one another, it is now worth looking at a particular case of the direct impact of media representations on people's understanding of genes.

The Impact of Media Representations on Understanding: Angelina Jolie's Double Mastectomy

On May 14, 2013, actress Angelina Jolie wrote a short essay for the *New York Times* in which she revealed that she had undergone a double mastectomy for preventive purposes. The reason was that she had been found to carry a "faulty gene," *BRCA1*, that "sharply" increased her "risk of developing breast cancer and ovarian cancer." In addition, her mother had died at the age of 56 after fighting for 10 years with cancer. Jolie noted that her doctors had estimated that her risk for breast cancer was 87 percent and her risk for ovarian cancer was 50 percent, as well as that "only a fraction of breast cancers result from an inherited gene mutation." She also noted that her chances of developing breast cancer after the double mastectomy dropped from 87 percent to 5 percent. She concluded her essay by writing:

> I choose not to keep my story private because there are many women who do not know that they might be living under the shadow of cancer. It is my hope that they, too, will be able to get gene tested, and that if they have a high risk they, too, will know that they have strong options. Life comes with many challenges. The ones that should not scare us are the ones we can take on and take control of.

In 2015, she wrote another essay in the *New York Times* in which she explained why she also had her ovaries and fallopian tubes surgically removed.

Past cases of celebrity health stories have been found to attract enormous media attention, to increase public awareness about the respective disease, and to impact public perception about it, as well as to influence health decisions. Therefore, it is interesting to look at what impact Angelina Jolie's essay had on laypeople's interest in and questions about breast cancer, how her choice to undergo double mastectomy was represented in the media, what people understood and inferred from these representations about breast cancer and its genetic basis, and whether there was any influence on people's decisions to undergo medical procedures related to breast cancer. The genes of interest in this case are *BRCA1* (breast cancer 1, on chromosome 17) and *BRCA2* (breast cancer 2, on chromosome 13) – humans have 23 pairs of chromosomes (22 autosomes, numbered 1–22, and a pair of sex chromosomes, either XX or XY). Particular *BRCA1* and *BRCA2* alleles have been associated with breast and ovarian cancers. *BRCA1* and *BRCA2* genes produce proteins that help repair damaged DNA and thus contribute to the stability of the genetic material of a cell by acting as tumor-suppressor genes. When certain mutations (changes in the genes) occur, DNA damage may not be repaired properly, and so cells are more likely to acquire additional mutations that may contribute to the development of cancer. It must be noted that mutation means "change," nothing more; whether this change has a good or a bad outcome is a different story. In the simplest case, there can be a change in a single "letter" in DNA. Such a mistake may not affect the message at all; it may affect it a little; or it may change it significantly (how mutations arise is explained in Chapter 2).

The publication of Jolie's essay aroused people's interest in breast cancer and exploded information searches about it. One study used digital surveillance of web data to analyze online search queries for breast cancer from 2010 to 2013 in the USA to investigate whether Angelina Jolie's announcement stimulated cancer-related information-seeking. It was found that, compared to the preceding six weeks, there was a 112 percent increase in general information queries, a 165 percent increase in risk-assessment queries, a 2,154 percent increase in genetics queries, and a 9,900 percent increase in treatment queries on the day of the announcement. Compared to the previous three years, both genetics and treatment search queries reached the highest levels ever. Another study measured the number of page views of the resources on the National Cancer Institute website over a nine-week period (three weeks before and six weeks after the announcement). The researchers also looked at the sources (including search engines and news outlets) used over the same period. It was found that overall there was a dramatic increase in page views on the day of the announcement: a 795-fold increase in page views of the "Preventive mastectomy" fact sheet; a 31-fold increase for the "*BRCA1* and *BRCA2*: cancer risk and genetic testing" fact sheet; and an 11-fold increase for the "Breast reconstruction after mastectomy" fact sheet. There was also a 5-fold increase for the "Genetics of breast and ovarian cancer" summary, which is intended for health professionals. Among the sources, news outlets had the largest impact. Whereas before the announcement news outlets comprised 0 percent of the sources, on the day of the announcement 86 percent of the referrals were from news outlets. Overall, these studies show the big impact that Jolie's announcement had on information-seeking about cancer.

So, many people looked for information on breast cancer following Jolie's announcement. But what did they find in the media? One study investigated the portrayal of Angelina Jolie's double mastectomy in the top five daily newspapers in Canada, the USA, and the UK. The sample included 103 newspaper articles published in the first month after the announcement (41 articles published in the USA, 34 in the UK, and 28 in Canada). About half of these articles were published during the first three days after the announcement, in the news section of those newspapers (not in the entertainment or lifestyle sections). The primary issue about the *BRCA* gene mutations that was highlighted in 72 of the articles was the increased risk of hereditary breast/ovarian cancer. Overall, 59 articles were supportive of Jolie's announcement, 33 presented it in a rather descriptive manner, and 6 included both positive and negative comments. Her decision was described as "brave and courageous" in 40 articles, as "rational, well-informed, and evidence based" in 23 articles, and as "empowering, inspiring, and a role model for other women" in 13 articles. At the same time, only 33 articles discussed that Jolie's condition was rare and that most women who get breast cancer are not carriers of *BRCA* alleles (this topic is discussed in detail in the next section). In addition, only 28 articles addressed the question of how effective the double mastectomy is compared to alternative methods, which were described as more effective in only 8 articles, whereas 19 articles mentioned its possible drawbacks. In conclusion, whereas newspaper articles were careful in how they presented medical information and did not make exaggerated claims about the efficacy of mastectomies, they were overall positive toward Jolie's decision.

Another study explored what types of information about Angelina Jolie's decision people could find on websites, as well as how that information was framed. The sample consisted of 92 open-access websites that had Jolie's decision as their main message. The majority of the websites briefly mentioned that Jolie had an extremely high risk for developing breast and ovarian cancer due to the *BRCA1* allele she carried. However, only a few of them explained when someone is susceptible, what the estimates associated with the various *BRCA* alleles are, and what factors put an individual at risk of developing cancer. The majority of the websites also briefly described Jolie's double mastectomy, focusing on the reasons for doing this and on the possible alternatives. In Jolie's case, the reasons for undergoing the mastectomy were her fear that she might die at a young age, like her mother, and her willingness to do as much as she could to avoid this. Many websites noted that not all health experts understood such a choice. In addition, 53 of them mentioned other alternatives to mastectomy. It should be noted that some of the websites mentioned Jolie's privileged position that allowed her to be able to afford the related medical procedures. All this information was either framed in a positive, negative, or a positive and negative way. Some websites described her decision as "empowering," "inspirational," or "brave," while noting that her public announcement was a "real service to women" and "helps others to learn and be informed." In contrast, other websites were concerned about the unintended consequences of her announcement, such as the overexaggeration of personal risk and the misperception that the prophylactic mastectomy was the only option.

All this brings us to the most important question: How did people perceive and understand this story, and what kinds of conclusions did they reach? An online survey conducted in the USA within three weeks of Jolie's announcement, with a representative sample of 2,572 adults, aimed at documenting whether participants recalled the Angelina Jolie story, what elements of the story they had retained, and how they understood it. The researchers found that 74 percent of participants were aware that Jolie had undergone a double mastectomy to reduce her risk of developing breast cancer. Among those participants who were aware of the story, 47 percent mentioned her risk as being 80–90 percent (the risk disclosed by Jolie herself was 87 percent). However, only 206 participants correctly reported the contribution of BRCA mutations to all breast cancer cases, and only 284 knew the risk for women without the BRCA alleles (with only 81 participants correctly answering both questions). Overall, 72 percent of participants thought that Angelina Jolie did the right thing to publicly announce her decision, with 73.5 percent of women and 65.7 percent of men thinking that she made the correct decision in undergoing the mastectomy. Interestingly, 57.4 percent of the women said that they would have undergone the surgery themselves if they carried the "faulty" gene, and 49.9 percent of the women said they would recommend the surgery to a family member in such a case. Overall, the authors of this study noted that whereas 3 out of 4 participants were aware of Jolie's story, fewer than 1 in 10 could correctly interpret the information about Jolie's risk of developing breast cancer relative to women unaffected by the BRCA gene mutation. In other words, an increased awareness of the story was not associated with an increased understanding of the condition.

A similar study in the USA was based on an online survey that was completed by 1,008 people, approximately six months after Jolie's announcement. The survey had questions about participants' awareness of the story, the genetic risk for breast cancer, their confidence applying their knowledge to a situation similar to Jolie's, and their views about prophylactic mastectomies. Participants had heard about Angelina Jolie's story on an average of 3.5 times. All genetic literacy skills questions were answered correctly by more than half of the participants, with 63 percent of them having at least four correct responses. In addition, 58 percent of the participants were at least somewhat confident in relying on their genomics knowledge to assess a mastectomy decision. About half of the participants (51 percent) neither agreed nor disagreed with whether women with BRCA mutations should have a mastectomy; among the rest, about 19 percent agreed and about 24 percent disagreed that women with BRCA mutations should have a mastectomy. Interestingly, the researchers found that high media exposure resulted in higher confidence about one's skills to assess such a situation for those who had been found to have lower genetic literacy skills. At the same time, they also found that increased confidence in applying genomics was associated with favoring mastectomies. The majority of the participants appeared to have appropriate confidence, as 70 percent of those with above-average confidence also had above-average genetic literacy skills. However, this was not the case for the remaining 30 percent. Therefore, repeated exposure to the media might make some people more confident than they should be, given their understanding of genetics, in making a decision such as the one Jolie made.

Overall, Angelina Jolie's decision was portrayed as a brave one in the media, and the majority of people thought that, given she carried the particular BRCA allele, she made the right choice to proceed to the prophylactic double mastectomy. However, the problem here is that a perceived genetic etiology was generally considered to justify the prophylactic mastectomy that Jolie underwent. But Jolie was not diagnosed to have cancer; rather, she was found to have an allele that was related to a higher probability than average for developing cancer. Without having any intention whatsoever to judge Jolie's decision, in the next section I consider in some detail the underlying facts, so that you get a better understanding of this situation.

Genetic Disease and Risk Literacy

Having presented the public representation and perception of the Angelina Jolie story, it is now useful to consider the underlying facts in a bit more detail. There are at least three important issues to consider. The first one is how people understand probabilities, and – perhaps most importantly – what probabilities mean. Earlier we saw that Angelina Jolie's risk for breast cancer was 87 percent. There are two types of risk: absolute and relative. Absolute risk is the probability for a person to develop a disease over a period of time, usually a lifetime (in this case, it is described as a lifetime risk). For example, an absolute risk of 0.4 indicates that the probability for a person to develop cancer over a period is 40 percent. In contrast, relative risk is defined with some reference point in mind. For instance, if one suggests that the relative risk for a person carrying a certain allele to develop a certain form of cancer is 1.5, this means that the probability of this person developing cancer is 50 percent higher than that of a person not having the same allele.

Psychologist Gerd Gigerenzer has shown convincingly that numbers like this may confuse people, and has suggested using frequencies rather than probabilities when talking about risks. An absolute risk of 0.4, which indicates that the probability for a person to develop the disease is 40 percent, practically means that 40 out of 100 people of a certain group will develop the disease over a certain period of time. Let us assume that this is a group of people who do not have a certain allele A. Imagine now that in another group of people carrying allele A, the absolute risk of developing the disease is 0.8. This means that 80 percent, or 80 out of 100 people, of this group will develop the disease. In this case, allele A is considered to increase the absolute risk for this second group by 40 percent. Now the relative risk is the ratio of the probability of people carrying allele A to develop the disease over the probability of people not carrying allele A to develop the disease. In this case, this would be $0.80/0.40 = 2$. A relative risk of 2 practically means that people carrying the allele have 100 percent higher probability, or twice the risk, to develop the disease than those who do not carry it (Table 1.4).

Given these considerations, what is the impact of the *BRCA* genes on the risk of development of breast cancer? Consider the following statements in the

	People who develop the disease	People who do not develop the disease	Total number of people	Absolute risk	Relative risk	Interpretation of relative risk
Group I: people without allele A	40	60	100	40/100 or 0.4	0.4/0.8 = 0.5	People in this group have 50 percent lower probability (half the risk) than people in group II to develop the disease
Group II: people with allele A	80	20	100	80/100 or 0.8	0.8/0.4 = 2	People in this group have 100 percent higher probability (twice the risk) than people in group I to develop the disease

Table 1.4 Absolute and relative risk

2019 report of the American Cancer Society: "Compared to women in the general population who have a 10 percent risk of developing breast cancer by 80 years of age, risk is estimated to be about 70 percent in women with pathogenic variants in *BRCA1* and *BRCA2*." What does this mean? It means that whereas by 80 years of age 10 out of 100 women of the general population will have developed breast cancer, this will be the case for about 70 out of 100 *BRCA1* and *BRCA2* mutation carriers. Obviously, the *BRCA* genes can make a big difference. But these probabilities also mean that one out of three women with these genes will not develop breast cancer. Most importantly, these data reveal nothing about whether these women will die from breast cancer.

This brings us to the second important issue to consider: Cancer is a complex disease that is not caused by a single gene. In other words, the mutated versions of *BRCA1* and *BRCA2* genes do not alone cause breast cancer. How cancer(s) develop is discussed in detail in Chapter 4. For now, it suffices to say that many genes contribute to the development of complex diseases such as cancer. For instance, one study followed 2,733 women aged 18–40 years who had been diagnosed with breast cancer. Among these, 338 women (12 percent) were found to have a pathogenic *BRCA* mutation (201 with *BRCA1*, 137 with *BRCA2*). A follow-up study of these women for up to 10 years showed that whereas there were 651 deaths due to breast cancer (out of a total of 678), there was no significant difference in overall survival between *BRCA*-positive and *BRCA*-negative women. In other words, women with young-onset breast cancer who carried a *BRCA* mutation had similar survival rates as noncarriers. This shows that the connection between the *BRCA* alleles and cancer is not deterministic but probabilistic.

The last issue to consider is how many women are actually concerned by the *BRCA* alleles. According to the American Cancer Society, the lifetime risk for a woman to develop breast cancer is 12.8 percent, or one in eight. This implies that over the course of a lifetime, on average one in eight women will develop breast cancer. Will those women die from it? Not necessarily, as the estimated lifetime risk for dying from breast cancer is estimated at 2.6 percent, or 1 in 39. Therefore, one should be careful when interpreting these probabilities. In addition, I must note that this is only statistics, not some law with predictive power. This means that it is possible that no woman within a particular group of 8 women will develop breast cancer and that no woman within a group of 39 people will die from it. Table 1.5 presents the estimated risk for women to develop breast cancer at various ages.

The report of the American Cancer Society also states:

> Inherited pathogenic (disease-causing) genetic variations in *BRCA1* and *BRCA2*, the most well-studied breast cancer susceptibility genes, account for 5 percent–10 percent of all female breast cancers and 15 percent–20 percent of all familial breast cancers. These variations are rare (about 1 in 400) in the general population, but occur slightly more often in certain ethnic or geographically isolated groups, such as those of Ashkenazi (Eastern European) Jewish descent (about 1 in 40).

This means that out of 100 women who will develop breast cancer, only 5–10 will be *BRCA1* and *BRCA2* mutation carriers. In other words, 90–95 out of 100 women who will develop breast cancer will not be *BRCA1* and *BRCA2*

Current age	Diagnosed with invasive breast cancer		Dying from breast cancer	
	%	n	%	n
20	0.1	1 in 1,479	<0.1	1 in 18,503
30	0.5	1 in 209	<0.1	1 in 2,016
40	1.5	1 in 65	0.2	1 in 645
50	2.4	1 in 42	0.3	1 in 310
60	3.5	1 in 28	0.5	1 in 193
70	4.1	1 in 25	0.8	1 in 132
80	3.0	1 in 33	1.0	1 in 101
Lifetime risk	12.8	1 in 8	2.6	1 in 39

Source: Data from American Cancer Society (2019). *Breast Cancer Facts & Figures 2019–2020.* Atlanta: American Cancer Society, table 2, p. 4.

Table 1.5   Age-specific 10-year probability of breast cancer diagnosis or death for US women

mutation carriers. Furthermore, if the *BRCA* alleles are rare, 1 in 400 or 0.25 percent, it means that very few women in the general population are affected. Therefore, Angelina Jolie's case is a rare one and most women should not be concerned.

Geneticist Mary-Claire King was awarded the 2014 *Lasker-Koshland Special Achievement Award in Medical Science* "for bold and imaginative contributions to medical science and society – exemplified by her discovery of a single gene BRCA1 that causes a … form of hereditary breast cancer." In a viewpoint article related to that award, King and her coauthors suggested that population-based screening of women for *BRCA1* and *BRCA2* should become a routine part of clinical practice. This should focus only on mutations that are clearly related to cancer development. The article concluded that:

> With modern genomics tools, all actionable mutations can be readily identified. Intensive monitoring and early invention protocols reduce risk in carriers. Sufficient knowledge is available to allow women to make informed decisions. Population-wide screening will require significant efforts to educate the public and to develop new counseling strategies, but this investment will both save women's lives and provide a model for other public health programs in genomic medicine. Women do not benefit by practices that "protect" them from information regarding their own health. They should have the choice to learn if they carry an actionable mutation in BRCA1 or BRCA2.

Attempting to confirm or disconfirm the presence of *BRCA* alleles might be important in the case of people with a family history of breast cancer, such as Angelina Jolie herself. In such cases, mutations can be considered as "actionable." But it should be made clear to people that genetic tests can only provide some information about the probability of developing a disease, and in no way can they predict whether this will happen or not. A woman who is overweight (a risk factor for breast cancer for older women) can lose weight, which would be beneficial for her health in general. But for a woman who finds out that she carries the *BRCA* alleles related to breast cancer, there is not much she can do. She could be one of the 70 out of 100 women (on average) who will develop breast cancer; but she could also be one of the 30 out of 100 women (again, on average) who will not develop breast cancer. The choice of double mastectomy that Angelina Jolie made is certainly an option. But whether this is the best one is a very personal decision. A woman carrying the *BRCA* alleles might not develop breast cancer, whereas an overweight woman who does not carry those alleles might. All medicine is probabilistic. Obviously, if people are well informed and they have understood the advantages and the disadvantages of the available options, then they could make their own decisions. What is more difficult to achieve is to realize, and live with, the underlying uncertainties.

The story is, in fact, even more complicated for various reasons. First, mutations in the *BRCA* genes produce a range of increases in cumulative average risk of breast and ovarian cancer up to a certain age. A "70 percent" risk is actually an average of a range of values. Second, there are different kinds of mutations in the *BRCA* genes, some of which increase risk more significantly than others. That is, there's not a single *BRCA1* or *BRCA2* mutation, but many different ones. Third, even for two individuals with the same mutation in the *BRCA* genes, their actual probability of getting cancer may not be the same because of the influence of other factors, such as diet, exercise, etc. Finally, surgery itself carries risk, so it should not be presented as a simple intervention that can save a woman from cancer.

Physicians have an important role in these decisions, as they might prefer to prescribe a genetic test in order to refrain from leaving themselves open to charges of negligence. But there is also the choice not to know. If one is going to develop a disease in *x* years (something that cannot be foretold) and live with the disease for another *y* years (how many, it is not possible to know), the *y* years living with the disease will likely be difficult independently of the values of *x* and *y*. Is there any point making the rest of one's life, the *x* years, stressful and miserable by anticipating the (probable or improbable) onset of the disease? Rather than live *x* stressful years and *y* years with the disease, one might choose not to know and to live in the present without worrying what the future will bring. Whether or not this is the best decision is subjective and debatable, but deciding not to know whether, for instance, a woman carries the *BRCA* mutations is a choice. Worse than this, a negative *BRCA* test result might make a woman falsely reassured that she will not develop breast cancer, but not having the specific alleles that a test looks for does not mean that a person does not have another one that the test simply did not detect.

Overall, about 15–20 percent of breast cancer is familial; that is, 15–20 percent of affected women have one or more first- or second-degree relatives with the disease. There exist several high-risk variants, which are very rare and which confer a relative risk of breast cancer higher than 4. These are variants in genes such as *BRCA1*, *BRCA2*, *TP53*, *STK11*, *CD1*, and *PTEN*, and they account for approximately 20 percent of the familial risk. There also exist moderate risk variants that confer a relative risk of breast cancer between 2 and 4, which account for up to 5 percent of the inherited familial risk. Finally, there exist more than 180 identified low-risk variants that confer a relative risk less than 1.5 and explain 18 percent of the familial risk. This entails that approximately 57 percent of the genetic background in familial breast cancer is unaccounted for now, and that there is more to breast cancer than the *BRCA1* and *BRCA2* genes. This is why a negative test does not tell us much.

I have thought a lot about these questions on a personal level. My maternal grandmother (who was overweight, by the way) died of breast cancer at the age of 58. My mother was 32 years old at the time. Nowadays, at the age of 65, she has never developed cancer, and has periodic tests to ensure that if cancer develops there will be time for treatment. Not knowing whether she carries an allele related to breast cancer has let her live her life without worrying. I do not think that a genetic test would have had any positive effect – in fact, a negative test might have falsely reassured her, made her believe she would never develop breast cancer and therefore neglect the precautionary periodic controls. I would not advise my own daughter to take a *BRCA* test either.

What is the conclusion? The extreme focus on *BRCA* mutations, especially through public accounts such as the Angelina Jolie story, perhaps distracts attention from other factors that affect more women than these alleles do. Women carrying the *BRCA* mutations should be concerned, but these are only a minority of the women who develop breast cancer. If there is a history of breast cancer in the family, then a woman should consider things differently than a woman without a family history. Of course, this way of considering probabilities makes no difference to someone who eventually develops the disease. However, the possible risks should be evaluated in a way that is as objective as possible. As sociologist Deborah Lynn Steinberg (who, perhaps ironically, died of breast cancer in 2017 at the age of 55) nicely put it: "As Jolie's decision and its wider public reception suggest, the persuasion of the singular gene paradigm is powerfully sedimented into public understanding of both cancer and genetics." A main aim of this book is to show that the singular gene paradigm is simply wrong for the vast majority of characteristics and disease.

Let us, then, begin our exploration of what genes are (and are not) and what genes do (and do not do).

2 The Origin and Evolution of the Gene Concept

Mendel and the Gene Concept

Perhaps you were taught at school that genetics began with Gregor Mendel. Because of his experiments with peas, Mendel is considered to be a pioneer of genetics and the person who discovered the laws of heredity. According to the model of "Mendelian inheritance," things are rather simple and straightforward with inherited characteristics. Some alleles are dominant – that is, they impose their effects on other alleles that are recessive. An individual who carries two recessive alleles exhibits the respective "recessive" characteristic, whereas a single dominant allele is sufficient for the "dominant" version of the characteristic to appear. In this sense, particular genes determine particular characteristics (e.g., seed color in peas), and particular alleles of those genes determine particular versions of the respective characteristics. Mendel, the story goes, discovered that characteristics are controlled by hereditary factors, the inheritance of which follows two laws: the law of segregation and the law of independent assortment.

In the first case, when two plants that differ in one characteristic, such as having seeds that are either round or wrinkled, are crossed, their offspring (generation 1) resemble one of the two parents (in this case, they have round seeds). In generation 2 (the offspring of the offspring) there is a constant ratio 3:1 between the round and the wrinkled shapes (Figure 2.1). Round shape is controlled by factor R, which is dominant, whereas wrinkled shape is controlled by factor r, which is recessive. Dominant and recessive practically

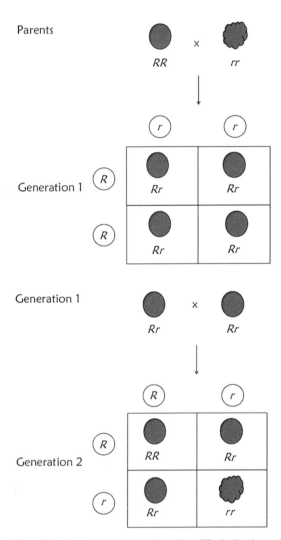

Figure 2.1   A cross between two plants that differ in the shape of seeds (round or wrinkled). Plants with round seeds have factors *RR* or *Rr*, whereas plants with wrinkled seeds have factors *rr*. The "wrinkled" characteristic "disappears" in generation 1 and "reappears" in generation 2 (all possible combinations of gametes are made).

means that when R and r are together, it is R that dominates over r and so the respective R phenotype is produced as if r was not even there. This means that plants with factors RR or Rr will have round seeds, whereas plants with rr will have wrinkled seeds. The explanation of these results is that the factors (R/r) controlling the different versions of the same characteristic (round/wrinkled) are separated (segregated) during fertilization and recombined in the offspring. This is described as Mendel's law of segregation.

When Mendel simultaneously studied the inheritance of two characteristics (e.g., both the shape of the seed and its color), he observed a similar but more complicated picture. When he crossed plants with yellow/round seeds and plants with green/wrinkled seeds, in generation 1 all offspring had yellow/round seeds. However, when those plants were crossed with each other, a constant ratio of 9 yellow/round: 3 yellow/wrinkled: 3 green/round: 1 green/wrinkled emerged in generation 2. This is the result of the combination of the probabilities to have all possible combinations of two characteristics – the one described earlier and a similar one regarding color (if you multiply 3:1 by 3:1 you get 9:3:3:1). Plants with factors YY or Yy have yellow seeds, whereas plants with yy have green seeds. The results suggested that the factors (R/r and Y/y) controlling the different characteristics (seed shape and seed color, respectively) were assorted independently during fertilization. As a result, all possible combinations were obtained (yellow/round, yellow/wrinkled, green/round, green/wrinkled), and this is why these are observed in generation 2 (Figure 2.2). This is described as Mendel's law of independent assortment.

In order to understand Mendel's work and its actual contribution, it is necessary to consider the historical context. Mendel was born in Moravia, which was a province of the Austrian Empire and is now part of the Czech Republic. Moravia was at that time a world-leading center of breeding practices. One such practice was hybridization, during which offspring were produced from the cross-fertilization of individuals of different species (for sexually reproducing organisms, a species can be defined as a group of, usually similar, organisms that can interbreed and produce fertile offspring). An important figure in these efforts was Cyril Napp, the abbot of the Augustinian Monastery of St. Thomas at Brno (Brünn). Napp accepted Mendel into the monastery in 1843, and also supported his studies at the University of Vienna from 1851 to

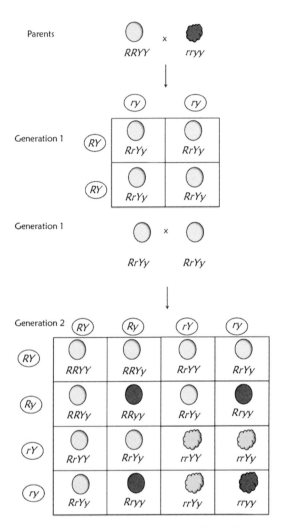

Figure 2.2   A cross between two plants that differ in the shape of seeds (round/wrinkled) and the color of seeds (yellow/green). In the second generation we find the ratio 9:3:3:1 (green peas appear here as having a darker color than yellow peas; all possible combinations of gametes are made).

1853. One of Mendel's teachers there was Franz Unger, who had argued that all plants were descended from common ancestors, thus rejecting the then prevalent idea that species do not change. Mendel probably learned from him about the hybridization experiments of Josef Gottlieb Kölreuter and Carl Friedrich von Gärtner, who had come to the conclusion that hybridization could not produce new species. This was in contrast to the view of the great taxonomist Carl Linnaeus, who had argued that hybridization could produce new species.

The question of whether or not hybridization could produce new species, and the broader understanding of this phenomenon, was of interest for the practical purpose of breeding related to agriculture and the socioeconomic context of Brno. Mendel's monastery was a center of learning in agriculture. In addition, the practical and financial benefits of new agricultural practices were a top priority for the monastery, as most of its income came from extensive land leasing to local farmers. It was in this context that Mendel began his hybridization experiments in 1856. He selected 34 distinct varieties of the edible pea (*Pisum sativum*) for his experiments, and subjected them to a two-year trial for purity in order to obtain varieties that when self-reproduced always produced plants with the same characteristics. Then he performed crosses between different varieties, focusing on seven characteristics, including the shape and the color of the seed.

Mendel's results were presented at the meetings of the Brno Natural Science Society in early 1865, and were published in the society's journal in 1866. At the beginning of his paper Mendel expressed his aim to study the development of hybrids in their descendants. He also noted that, until that time, no one had succeeded in establishing a generally valid law for the formation and development of hybrids. Mendel described the transmission of characteristics over generations bred from hybrids. In particular, he observed that the hybrids obtained from the various crosses between different varieties were not always intermediate between the parental forms. Rather, some hybrids exhibited certain characteristics exactly as they appeared in the parental plants. Mendel called dominant the parental characteristics that also appeared in the hybrids, and recessive the parental characteristics that did not appear in the hybrids but reappeared fully formed in the next generation. Therefore, Mendel studied the transmission of characteristics, not of hereditary particles, across generations; he did not "discover" genes! He also studied hybridization in particular, not heredity in general, and it should therefore be no surprise that the word "heredity" does not appear in his paper.

The term "heredity" in the modern, biological sense – that is, with reference to the transmission of some substance across generations – does not appear in writings on the generation of organisms until the mid-eighteenth century. The systematic use of this term was introduced around 1800 by French physicians, and was soon incorporated into other European languages. The term "heredity" derives from the Latin *hereditas*, which means inheritance of succession. The biological concept of heredity resulted from the metaphorical use of a juridical concept, which referred to the distribution of status, property, and other goods according to a system of rules about how these should be passed on to other people once the proprietor passed away. Today "heredity" is considered to be a biological concept, whereas "inheritance" is used both in biological and nonbiological contexts. In the present book, "genetic inheritance" refers to the process of transmission of genetic material across generations, whereas "heredity" refers to the broader phenomenon of which this process is part.

References to heredity are found in Herbert Spencer's *Principles of Biology,* published in 1864. At that time, the mechanism of heredity was at the center of biological thought, in part because Charles Darwin's theory of descent with modification through natural selection (published in 1859 in the *Origin of Species*) lacked a theory that could explain the origin and inheritance of new variations (i.e., differences) that were so central to it. In response to this problem, Darwin proposed in 1868 his *Provisional Hypothesis of Pangenesis*, which literally means "origin from everywhere." According to this hypothesis, all parts of the body contributed to the formation of the offspring by producing microscopic entities, the gemmules, which somehow carried the organismal properties from generation to generation. A central idea was also that the inheritance of acquired characteristics was possible – that is, that characteristics acquired by the parents during their lives could be passed on to their offspring.

In 1871, Francis Galton tested experimentally the hypothesis of pangenesis and practically disconfirmed it. In 1876, he proposed his own theory of heredity, suggesting that hereditary factors did not arise from the various body tissues. He also proposed the term "stirp" that accounted for the total of the hereditary elements or germs at the fertilized ovum, thus postulating a form of germline theory – that is, that reproductive cells existed separately from body (somatic) cells. Galton also suggested that evolutionary change might take place in a discontinuous manner and not gradually. William Keith Brooks took this idea further in 1883 to suggest that evolution proceeded with extensive modifications, and not gradually as Darwin had suggested. Carl von Nägeli proposed a theory that was based on the existence of hereditary factors. Hugo de Vries retained some of Darwin's ideas, and by combining breeding experiments with Galton's statistical methods, he suggested in 1889 a modified theory of pangenesis. In the meantime, during the 1880s, vital dyes and improved microscopes made possible the visualization of cellular structures and processes. It was thus shown that reproductive cells existed independently of the other body tissues. August Weismann drew on these new findings to propose that characteristics were inherited only from the germline, not from the whole body.

All these scholars were aware of one another's work and worked, actively and interactively, to develop a theory of heredity. Mendel is nowhere in this picture. Only Nägeli came to know of Mendel's experimental work through their correspondence from 1866 to 1873. Following Nägeli's advice, Mendel worked on *Hieracium* (hawkweed, a genus of the sunflower family) from 1866 to 1871, which gave him different results from those of *Pisum*. Nägeli did not seem to pay much attention to Mendel's work, but on at least one occasion he cited Mendel's 1866 paper. Most importantly, the Brno Natural Science Society sent more than 100 copies of the journal that included Mendel's paper to scientific centers around the world. At least 10 references to Mendel's paper appeared in the scientific literature before 1900, some of them in books that were widely read by naturalists. However, Mendel's work did not become widely known, probably because it was not an explicit attempt to develop a theory of heredity that was of interest to naturalists at that time. Mendel was rather interested in understanding hybridization and its patterns, which would be of practical, agricultural interest. It was in this practical, local context that Mendel's work made sense in his time.

During the latter half of the nineteenth century, Galton and Weismann developed frameworks of "hard" heredity – that is, one characterized by discontinuous variation and nonblending characteristics. In addition, Galton postulated and Weismann established the idea of the germline – that is, that reproductive cells exist independently of the other cells of the body. In the late 1870s, Walther Flemming observed and described mitosis, and Oscar Hertwig observed and described meiosis (discussed in the next paragraph). In the 1880s, it was found that fertilization involved the fusion of two nuclei, and how this occurred at the level of chromosomes. All these developments provided a new context for reading Mendel's paper that brought him back into the scene in 1900, when Hugo de Vries, Carl Correns, and Erich von Tschermak published the results of their research on plant hybridization that agreed with those of Mendel. Read in a new context, Mendel's paper was considered as bringing together the findings of breeding experiments and of microscopic studies of cells, showing that particulate hereditary factors existing in the nuclei of cells were segregated and independently assorted.

Let us briefly revisit some basic biology here. Mitosis is the division of the cell nucleus that finally results in two cells having exactly the same number of chromosomes as each other and as the initial cell. This is how somatic cells proliferate. Meiosis is the division of the nucleus that finally results in four sperm cells (but only one ovum, as the other cells become polar bodies) having half the number of chromosomes of the initial cell. This is how reproductive cells are formed. It is due to meiosis that organisms consisting of diploid cells (i.e., cells that have pairs of chromosomes) produce haploid gametes (i.e., reproductive cells that have one pair from each chromosome). This is why in Figures 2.1 and 2.2 only one of the factors R or r or only one of the factors Y or y were included in the gametes. Meiosis is important because in this way diploid cells produce haploid gametes that, when they fuse, result in diploid embryos. Each of the chromosomes of the same pair is inherited from each one of the parental organisms (Figure 2.3). It should be noted that DNA is duplicated before mitosis and meiosis.

After 1900, the work of Mendel guided the development of the new science of "genetics," a term coined by William Bateson in 1906. Bateson's book *Mendel's Principles of Heredity: A Defence* contains the first English

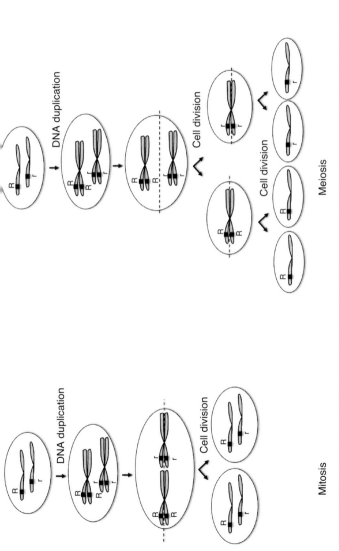

Figure 2.3 Mitosis and meiosis. Mitosis results in two cells with the same number of chromosomes as the original cell, whereas meiosis results in four cells with half the number of chromosomes of the original cell.

translation of Mendel's paper. In this book, Bateson presented Mendel's work as providing the solutions for various problems relevant to heredity. He also introduced new terms such as "allelomorph," "heterozygote," and "homozygote." Allelomorph referred to the different versions of the same characteristic and to the respective hereditary factors. In 1927, George Shull introduced the shorter term "allele," which is the common term used today to refer to the alternative versions of the same gene. Currently, humans and other organisms are considered to carry pairs of alleles; an individual carrying the same allele twice is described as a homozygote, whereas an individual carrying two different alleles is described as a heterozygote.

Already in 1902, however, Raphael Weldon showed that Mendel's "laws" might not actually work even for peas. His studies of varieties of pea hybrids led him to conclude that there was a continuum of colors from greenish yellow to yellowish green, as well as a continuum of shapes from smooth to wrinkled. It thus appeared that in obtaining purebred plants for his experiments, Mendel had actually eliminated all natural variation in his peas, and that therefore characteristics were not usually as discontinuous – for instance, either yellow or green – as in the varieties he worked with. William Castle, and later Bateson and his colleagues, also found deviations from Mendel's ratios and thus realized that these were not universal. But Mendel's methods worked too well to be abandoned. One reason was that they facilitated understanding of the role of chromosomes in heredity. In 1903, Walter Sutton performed a careful study of the process of cell division and concluded that a large number of different combinations of maternal and paternal chromosomes were possible in the mature reproductive cells of an individual. This evidence provided the foundations for explaining the physical basis of Mendel's ratios and for understanding the chromosomal nature of heredity.

Eventually, it was Wilhelm Johannsen who, in 1909, proposed the term "gene." Etymologically, the term derives from the hereditary factors of de Vries' *(Intracellular) Pangenesis*, which were called "Pangens" and were occasionally transcribed as "Pangenes," whereas the idea goes back to Darwin's *Pangenesis*. Johannsen also noted that: "The word gene is completely free from any hypothesis; it only expresses the established fact, that at least many properties of an organism are conditioned by special, separable and thus independent 'conditions', 'foundations', 'dispositions'." Johannsen also coined the terms "genotype" and "phenotype." In today's terms, the genotype refers to the alleles related to a particular characteristic that an individual has, whereas the phenotype refers to the outcome of development with reference to a particular characteristic. Johannsen considered genes as real entities and contrasted them to the speculative hereditary particles proposed in earlier theories. However, for him the gene concept was free from any assumption about its localization in the cell and its material constitution.

The "Classical" Gene

Many answers to questions about genes were given by an influential research program led by Thomas Hunt Morgan and his collaborators at Columbia University. This program also established *Drosophila* (fruit flies) as a model organism in genetics research. As early as 1913, Morgan tried to clarify the relation between "unit characters" and "unit factors" – he had not yet adopted the term "gene": "The confusion is due to a tendency, sometimes unintentional, to speak of a unit character as the product of a particular unit factor acting alone, but this identification has no real basis." Using several examples, Morgan explained that to think that one factor alone could determine a characteristic was misleading and stemmed from a misunderstanding of the notations used. It might be a surprise for you to read that the simplistic notion of factors (genes in today's parlance) that alone determine phenotypic characteristics – or "genes for" – which is still taught in some biology classes today and which – as we saw in Chapter 1 – exists in various forms in the media, was rejected by researchers in genetics more than 100 years ago.

Consider the following excerpt from a classic book written by Morgan and his students:

> Mendelian heredity has taught us that the germ cells must contain many factors that affect the same character. Red eye color in Drosophila, for example, must be due to a large number of factors, for as many as 25 mutations for eye color at different loci [positions on chromosomes] have already come to light … Each such color may be the product of 25 factors (probably of many more) and each set of 25 or more differs

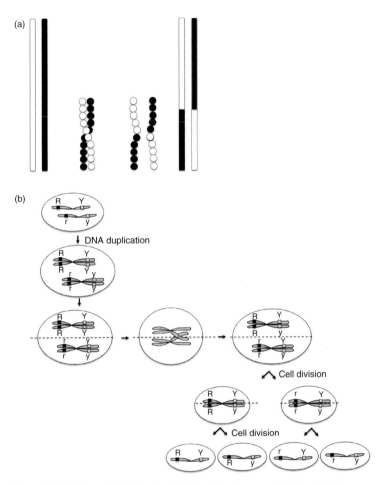

Figure 2.4 (a) A figure like this in the 1915 book by Morgan and his colleagues represented chromosomes like strings consisting of beads, which were the genes, called unit factors at that time. It also depicted how crossing over took place and resulted in different combinations of genes (e.g., "black and white" instead of "black only" or "white only") (based on Morgan et al.'s *The Mechanism of Mendelian*

from the normal in a different factor. It is this one different factor that we regard as the "unit factor" for this particular effect, but obviously it is only one of the 25 unit factors that are producing this effect … The converse relation is also true, namely, that a single factor may affect more than one character … Failure to realize the importance of these two points, namely, that a single factor may have several effects, and that a single character may depend on many factors, has led to much confusion.

This is the important distinction between *genes as characteristic-makers* and genes as *characteristic-difference-makers*: it is one thing to suggest that a gene alone makes a characteristic, and another that a gene makes a difference in the state of a characteristic (this is discussed in detail in Chapter 5). More than 100 years ago, it was clear to researchers that there is no single *gene–characteristic* connection, as a single gene might affect various characteristics and many genes might affect a single characteristic. Beyond that, though, they understood neither how genes affected characteristics, nor what exactly their effects on these characteristics were. It should be noted, at this point, that the idea that there is a relation between a difference in a gene and a difference in a characteristic allows disregarding environmental influences. The reason for this is that the research by Morgan and his colleagues was conducted in controlled laboratory conditions, where environmental changes were slight and could thus be ignored.

Another important idea in that book by Morgan and his colleagues was that genes are located sequentially on chromosomes, like beads on a string (Figure 2.4a). Figure 2.4a is indicative of how genes were conceived as discrete parts of chromosomes. Morgan and his collaborators had realized that several genes were not inherited independently; in contrast, there was some kind of genetic linkage between them that in turn pointed to a physical linkage. In this sense,

---

Caption for Figure 2.4 (cont.)

Heredity, p. 60). (b) How crossing over of homologous chromosomes occurs during meiosis. This exchange of chromosome parts creates combinations of alleles (in this case Ry and rY) that did not exist in the parents.

genes could be conceived as beads on the same string. This conceptualization was essential for the techniques used by Morgan and his colleagues for mapping genes, and for the process of crossing over depicted in Figure 2.4a. Crossing over is the phenomenon of exchange of chromosome parts between two homologous chromosomes during meiosis (Figure 2.4b), which results in new combinations of genes in offspring that did not exist in their parents.

The first genetic map – that is, the first map showing the linear arrangement of genes on chromosomes – had been published in 1913 by Alfred Sturtevant, a student of Morgan. In that paper, Sturtevant also set out the logic for genetic mapping. The number of crossovers per 100 cases was used as an index of the distance between any two genes (still described as factors in that paper). If one could thus determine the distances between genes A and B and between genes B and C, one would also be able to predict the distance between genes A and C. Therefore, the relative positions of genes could be empirically mapped on chromosomes, and this gave them a more material character than before. This understanding became possible at that time in part because of luck. Morgan and his collaborators worked with *Drosophila*, which has only four chromosomes, and this made more probable the identification of genes that were linked than if the organism had 46 chromosomes, as we do.

Morgan and his colleagues dominated the field described as classical genetics. It was due to their adoption of the term "gene" that it became widely used. Morgan used the term for the first time in 1917: "The germ plasm must, therefore, be made up of independent elements of some kind. It is these elements that we call genetic factors or more briefly genes. This evidence teaches us nothing further about the nature of the postulated genes, or of their location in the germ plasm." Then Morgan asked the question: "Why, it may be asked, is it not simpler to deal with the characters themselves, as in fact Mendel did, rather than introduce an imaginary entity, the gene." This quotation is interesting in itself, as it indicates that Morgan thought that Mendel had worked with characteristics only (as I argued in the previous section) and not that he had discovered genes. Morgan nevertheless defended the use of the concept of genes, even if there was no evidence about their nature and localization. The reason for this was that the gene concept had already been a very valuable heuristic tool for conducting research.

Morgan and his colleagues studied several characteristics in *Drosophila*, such as eye color, which were related to sex chromosomes (X and Y) and so exhibited different ratios than those observed by Mendel. Ratios like 3:1 and 9:3:3:1 were observed for genes located on autosomes, the chromosomes found in both sexes. However, when genes were located on sex chromosomes these ratios changed. The reason for this is that, whereas in organisms like *Drosophila* there always exist two alleles for a gene located on autosomes, there are some genes located on X chromosomes for which there is only one allele in males as they only have one X chromosome. In this case, assuming that there exist two alleles (e.g., X^A and X^a), females can be homozygous ($X^A X^A$ or $X^a X^a$) or heterozygous ($X^A X^a$), whereas males are hemizygous (either $X^A Y$ or $X^a Y$). As shown in Figure 2.5, whether it is the female parent or the male parent that has white eyes makes a difference in which characteristics the offspring will have.

A major contribution summarizing the conclusions from the work of Morgan's group was his 1926 book *The Theory of the Gene*, in which he presented all the available evidence showing that genes were located on chromosomes. According to him, the most complete and convincing evidence for the importance of chromosomes in heredity came from research that showed the specific effects of particular changes in the number of chromosomes. It had been observed that flies lacking one chromosome 4 developed to be slightly different in many parts of the body compared to the "normal" ones. These results showed that the presence of a single chromosome 4 was not adequate to produce a "normal" phenotype, and that therefore some crucial genes should be located on the missing one. Morgan described the theory of the gene as one that was based on particular principles: (1) that the paired genes related to the characteristics of an individual; (2) that the genes of the same pair separated in accordance with Mendel's first law; (3) that genes that were not linked assorted independently in accordance with Mendel's second law; (4) that crossing over took place; and (5) that crossing over provided evidence for the linear order of genes and their relative positions. Morgan's theory was also based on certain assumptions: that genes were relatively constant, that they could be multiplied, and that they were united and then separated during the maturation of the reproductive cells. He concluded the book by estimating that genes should have the

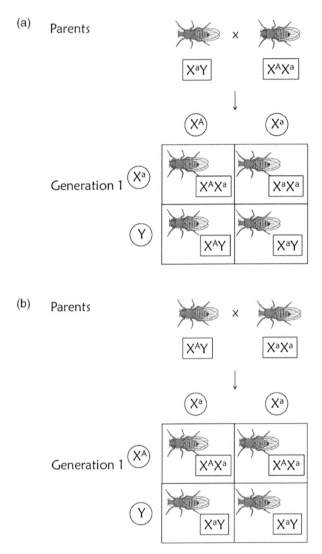

Figure 2.5   X-linked inheritance was studied by Morgan and his colleagues. When the male has white eyes and the female has red eyes (being darker than white eyes in this

size of the larger organic molecules, and that they were constant exactly because they were chemical entities.

In 1927 Hermann J. Muller, one of Morgan's students, provided crucial evidence that genes could be chemical entities. He reported that treatment of sperm with relatively heavy doses of X-rays caused an increase in mutation rate in *Drosophila* of about 15,000 percent in the treatment group compared to the control group. What was more interesting was that the characteristics of the mutations produced by X-rays were generally similar to those previously observed in *Drosophila*. Actually, most of the already known mutant phenotypes were produced during the experiments. Muller noted that the interest of these experiments lay "in their bearing on the problems of the composition and behavior of chromosomes and genes." That his experiments produced most of the mutant phenotypes already observed in *Drosophila* suggested that genes had some material composition that was altered by X-rays. This provided crucial evidence that genes should be material entities of some kind.

Additional crucial evidence that genes were material entities came from studies of crossing over. This phenomenon was taken for granted as early as 1915, but it was only in 1931 that Barbara McClintock and her colleague Harriet Creighton showed that there was a correspondence between chromosomes exchanging parts and the recombination of phenotypes, and therefore that genes were located on chromosomes. Their aim was to show that cytological crossing over occurs and that it is accompanied by genetic crossing over. McClintock had found that in a certain strain of corn, chromosome 9 had "a conspicuous knob at the end of the short arm." Therefore, to show the correlation between cytological and genetic crossing over, it was necessary to have plants that differed in the presence of the knob and of

---

Caption for Figure 2.5 (cont.)

figure), half of the male and half of the female offspring will have white eyes, whereas the other halves will have red eyes (a). But when it is the male that has red eyes and the female that has white eyes, all the female offspring will have red eyes and all the male offspring will have white eyes (b).

particular linked genes. By crossing plants that differed appropriately in these characteristics, it was shown that cytological crossing over occurred and that it was accompanied by the expected types of genetic crossing over. It was thus concluded that chromosomes of the same pair exchanged parts at the same time that they exchanged genes assumed to be located on these regions.

Morgan received the 1933 Nobel Prize in Physiology and Medicine "for his discoveries concerning the role played by the chromosome in heredity." In his Nobel lecture, Morgan suggested that simple Mendelian genetics could not account for the development of phenotypes as "the gene generally produces more than one visible effect on the individual, and ... there may be also many invisible effects of the same gene." In the same lecture he also clearly explained why at that point it did not make much difference for geneticists to be aware of what genes were made of: "There is no consensus of opinion amongst geneticists as to what the genes are – whether they are real or purely fictitious – because at the level at which the genetic experiments lie, it does not make the slightest difference whether the gene is a hypothetical unit, or whether the gene is a material particle." Therefore, the gene was still a hypothetical entity, a conceptual tool with a heuristic value. This gene concept, which I call the "classical" gene hereafter, was an extremely useful tool for doing research – genetic analysis as it is called – but its explanatory potential was relatively limited.

The "Molecular" Gene

Until the 1940s, the dominant research approach had been to study the effect of mutations on organisms and then work out the details of the respective biochemical processes. But in 1941, George Beadle, a former student of Morgan, and his colleague Edward Tatum came up with a new idea: Instead of understanding the biochemistry of gene products, they thought they might start with well-known metabolic compounds and work out their genetics. As *Drosophila* was a rather complex organism for this kind of study, they turned to a much simpler organism, the bread mold *Neurospora crassa*. In 1941, they published their results, concluding that a single gene seemed to somehow correspond to a single protein in a certain metabolic pathway. They found that the inability in mutant *Neurospora* strains to synthesize particular molecules, such as vitamins B1 and B6, was "inherited as though differentiated from normal by single genes." They showed that different *Neurospora* strains, requiring one or the other vitamin to survive, were the outcome of mutations in single genes. This in turn showed that genes somehow regulated biochemical processes and the synthesis of enzymes. Beadle and Tatum suggested that the relation between genotype and phenotype could thereafter be conceptualized in terms of gene and gene product, and so it was possible to describe the effects of genes in more concrete, molecular terms. This relation became widely known as the "one gene–one enzyme hypothesis." Yet, what genes were made of remained unknown.

Around the same time, Oswald Avery, Colin MacLeod, and Maclyn McCarty published experimental results suggesting that DNA was the hereditary material. In their experiments, it was found that DNA was the material that transformed bacteria of one type to another. This conclusion was not widely accepted, because at the time many scientists believed that only proteins, not DNA, had the necessary specificity to be genes. One reason for this is that proteins consist of 20 different amino acids, whereas DNA consists of 4 different nucleotides. Therefore, for a molecule that consists of 100 subunits, there can be 20^{100} different combinations if it is a protein and 4^{100} (a lot less) if it is DNA. Proteins were thus considered as more variable than DNA, and therefore as more capable of having the necessary specificity to be the genetic material. Eventually, in 1952, Alfred Hershey and Martha Chase showed that most of the material from a bacteriophage (a virus that infects bacteria) entering a bacterium was nucleic acid and not protein. This made many researchers accept that DNA was the genetic material.

The landmark year for the elucidation of the nature and structure of genes is generally considered to be 1953, when James Watson and Francis Crick proposed a model for the structure of DNA. They performed no experiments for this purpose. Rather, they adopted the model-building approach of Linus Pauling, a prominent chemist of the time, in order to figure out the structure of DNA by relying on experimental data provided by others. Erwin Chargaff had shown that any DNA molecule contained equal proportions of adenine and thymine, as well as of guanine and cytosine. John Griffith had pointed out that adenine and thymine, as well as guanine and cytosine, could fit together, linked by hydrogen bonds. Maurice Wilkins and Rosalind Franklin had performed X-ray diffraction studies of DNA, and their results suggested a spiral structure for the molecule. Watson and Crick built actual models on the basis of these data that helped them arrive at the double-helix structure of DNA. Based on all of this, in April 1953 Watson and Crick famously proposed the double-helix model (Figure 2.6) for the structure of DNA. The same issue of *Nature* also included two papers that provided the evidence for this model. One of these, coauthored by Wilkins, analyzed the evidence from X-ray diffraction to show that the structure was indeed helical and that it existed in a natural state. The other paper was coauthored by Franklin, and provided further support for the "probably helical" structure model suggested by Watson and Crick.

Let us briefly consider the details of the double-helix model of DNA. A DNA molecule consists of two nucleotide strands linked to each other. Each of these strands in turn consists of nucleotides, which differ only in their constitution of each of four "bases": adenine (A), thymine (T), cytosine (C), and guanine (G). The nucleotides of the same strand are connected by relatively strong bonds in a linear sequence. The nucleotides of the opposite strands are connected through relatively weak bonds that are formed always between an A and a T or a C and a G. There exist two bonds within each A–T pair and three bonds within each G–C pair. In their article, Watson and Crick noted that "the specific pairing we have postulated immediately suggests a possible copying mechanism for the genetic material." This simply meant that, as there were only two possible pairs of nucleotides (A–T and G–C), the two strands would be complementary – that is, if there was a certain base at a certain point of one strand, then the other base of the A–T/G–C pair should be on the respective point on the other strand. For example, if the sequence in one strand were GATTACA, then it could be easily inferred that the sequence in the respective position in the other strand would be CTAATGT (because A pairs with T, and G with C).

Significant support for the double-helix structure of DNA became available a few years later when Matthew Meselson and Franklin Stahl showed that each of the two new DNA molecules produced during DNA replication consisted of one strand from the initial DNA molecule and a newly synthesized one. This suggested that when a DNA molecule was replicated, each strand served as a template for the synthesis of a complementary one. For example, a DNA molecule consisting of an GGTGTT and a CCACAA strand would produce two molecules after replication. One of them would consist of the initial GGTGTT

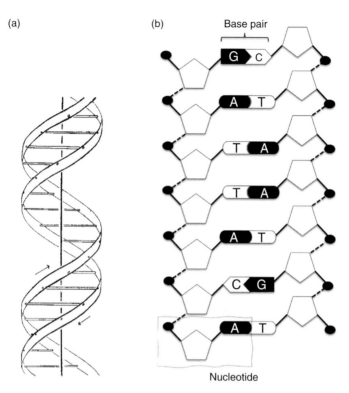

Figure 2.6   (a) The double-helix structure of DNA as imagined by Watson and Crick and as drawn by Crick's wife, Odile, for their *Nature* paper. Reprinted by permission from Macmillan Publishers Ltd: *Nature* (pp. 171, 737–738), © 1953. (b) A representation of the structure of a DNA molecule. All nucleotides have the same basic structure and differ in which of the four bases A, T, G, or C, they contain. Note that A and G are similar, and so are T and C. A is complementary to T, whereas G is complementary to C. The resulting DNA molecule consists of two complementary and antiparallel nucleotide strands.

strand and a new CCACAA one. The other molecule would consist of the initial CCACAA strand and a new GGTGTT one (Figure 2.7a). It should be noted that the term "self-duplication" used by Watson and Crick in their article, as quoted

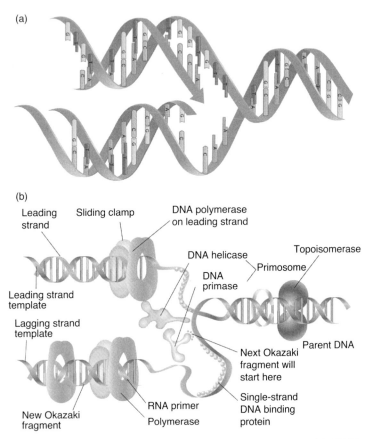

(a)

(b)

Leading strand

Sliding clamp

DNA polymerase on leading strand

DNA helicase

Topoisomerase

Primosome

DNA primase

Leading strand template

Lagging strand template

Parent DNA

Next Okazaki fragment will start here

New Okazaki fragment

RNA primer

Polymerase

Single-strand DNA binding protein

Figure 2.7   (a) An overview of DNA replication, in which each strand of the initial DNA molecule serves as a template for the synthesis of the complementary strand. Thus, two new molecules emerge, each consisting of an old and a newly synthesized strand. These two new molecules would be identical if there were no mutations. This is the portrayal of DNA as a self-replicating molecule, which can replicate on its own and produce two identical molecules (© ellepigrafica). (b) A more detailed and more accurate representation of the process of DNA replication, which includes some of the various proteins involved in the process. This portrayal shows that DNA is not a self-

earlier, and the general description of the process of DNA replication as "self-replication," has supported reference to DNA as a self-replicating molecule. However, this is not the case. The process of DNA replication depends on various proteins, such as DNA helicases and DNA polymerases (Figure 2.7b), needed for opening and copying the molecule, as well as for correcting various kinds of mistakes that occur accidentally during replication or due to mutagens.

There are several kinds of mistakes that are possible during DNA replication. One example is the change from a G–C pair to an A–T one. The base G normally occurs in the "keto" form (H–N–C=O); however, it can occasionally change to the less stable "enol" form (N=C–OH). The latter form of G pairs with T rather than with C. As a result, during DNA replication it is a T and not a C that is inserted opposite an enol-G. At the next round of replication T will be paired with A, thus producing a permanent change in the DNA sequence, as the pair in that position will thereafter be A–T and not G–C, as it was in the initial molecule. Changes in DNA sequences can also occur because of mutagens (i.e., factors directly affecting the structure of DNA). One example is UV light, which causes adjacent Ts to join. When these are later excised and repaired by repair enzymes, there is a possibility of error because other bases might be put in their place. Such changes in DNA sequences are described as mutations when they bring about a significant phenotypic change. Generally speaking, we can distinguish between germline mutations – that is, those that occur in any germ (reproductive) cell of the body and that we usually inherit from our parents – and somatic mutations – that is, those that occur in any cell of our body (except for the germ cells) after conception, such as those that initiate the development of cancers. If such changes do not bring about any changes, they are often described as variants or polymorphisms.

It was during the 1950s that the first evidence became available that mutations in genes affecting the structure of proteins are related to disease. Already

---

Caption for Figure 2.7 (cont.)

replicating molecule because it depends on various proteins for its replication. This is very important as DNA replication can take place only when these proteins are present (either *in vivo* – that is, within cells – or *in vitro* – that is, in the test tube, such as during polymerase chain reaction) (© snapgalleria).

in 1949, Pauling had shown that sickle cell anemia, a disease in which red blood cells become sickle-shaped, was due to an abnormal structure of hemoglobin, the protein that transfers oxygen through blood to the various tissues. In the same year, James Neel had also shown that sickle cell anemia was inherited as a Mendelian characteristic. However, it took a few more years for the connection between the inheritance of a disease and its molecular basis – a specific change in the structure of a protein – to be established. It was eventually in 1956 that Vernon Ingram showed that the formation of hemoglobin found in patients with sickle cell anemia was due to a mutation in a single gene. This showed how a mutation in a gene could cause a change in a single amino acid, which in turn could have a major impact on the structure of a protein and thus on the emergence of a disease.

Around the same time the conceptualization of genes began to change. The classical genetics conceptualization of genes as beads on a string (Figure 2.4a) implied that genes were indivisible. By the early 1950s, the gene had come to be perceived in various distinct ways: as the unit of physiological function, as the unit of mutation, or as the unit of recombination via crossing over. However, it had already been shown that these different kinds of units did not always converge on the same entity. Influenced by this conclusion, and by the then-recent proposal of the structure of DNA by Watson and Crick, Seymour Benzer thought that the size of genes could be estimated. He decided to conduct experiments with bacteriophages and he concluded that it was possible to resolve the structure of genes, perhaps even at the level of a single nucleotide. The question that Benzer aimed to answer was whether particular mutations were part of the same functional unit or not. His results showed that the unit of function was larger than both the unit of mutation and the unit of recombination. This meant that mutation and recombination *within* a gene were possible. The major conclusion he thus made was that genes were divisible and that they consisted of nucleotides. Benzer's major contribution was therefore to show how there could be a correspondence between a gene and a sequence of nucleotides on DNA. This was the foundation for the "molecular" gene concept.

Based on Benzer's conclusions, Crick then proposed "the sequence hypothesis," which assumed that the sequence of the bases of a nucleic acid encodes the amino acid sequence of a protein. This served to characterize the relationship between DNA, RNA, and proteins. An important contribution made by Crick was the introduction of the metaphor of information, defined as the precise determination of sequence, either of bases in the nucleic acid or of amino acids in the protein. Thus, DNA, RNA, and proteins were related to one another based on the information encoded within their sequence. These ideas were the foundation for Crick's "central dogma": The transfer of information, in the sense of the precise determination of sequence (either of nucleotides in nucleic acids, or amino acids in proteins), might be possible from nucleic acid to nucleic acid, or from nucleic acid to protein, but not from proteins to nucleic acids. The "molecular" gene was thus conceived as a sequence of DNA with a specific structure and particular properties, encoding sequence information for the synthesis of a protein.

The next big question was how the information encoded in DNA was "expressed," leading to the production of an RNA that might further be translated into a protein, a process described as gene expression. A crucial step in answering this question was made in 1961 through the work of François Jacob and Jacques Monod, which revealed the role of messenger RNA (mRNA). Already in 1955, a connection had been shown between a type of RNA (later called ribosomal RNA or rRNA) and some small structures that were considered as the platform for protein synthesis, called ribosomes. Then, it was also concluded that another type of RNA molecule (transfer RNA or tRNA) was essential for protein synthesis. RNA is a molecule quite similar to DNA, but with some differences, including that RNA is a single-stranded molecule and that it contains uracil (U) instead of thymine (T).

All these findings helped Crick produce a new concept of biological specificity that was based on the transfer of sequence information from one type of macromolecule to another. The first to propose an account for this biological specificity and the transfer of information was the physicist George Gamow, in 1954. He suggested that the DNA molecule itself could serve as a template for protein synthesis by proposing a one-to-one correspondence between the 20 independent combinations of consecutive triplets (with each triplet sharing two bases with each of the adjacent ones) and the 20 amino acids found in organisms. Interestingly, this code yielded exactly 20 triplets that could correspond to the 20 amino acids. But this is not how it works.

In 1957, Crick was the lead author of an article that suggested a structure for the genetic code. Given that there were four different nucleotides in DNA (with A, T, C, or G), and that 20 amino acids were found in the proteins of organisms, it was concluded that a single nucleotide could not correspond to a single amino acid, as only four amino acids would thus be coded. In addition, two nucleotides for one amino acid would give $4 \times 4 = 16$ combinations, which would still not be enough. Therefore, three nucleotides should correspond to one amino acid. Each of the $4 \times 4 \times 4 = 64$ possible triplets of nucleotides should somehow correspond to each of the 20 amino acids. In 1961, Crick, Sydney Brenner, and two other colleagues drew on experimental evidence to propose several features of the genetic code: that triplets of nucleotides coded one amino acid (they considered, however, that a triplet consisting of the same nucleotide could code one amino acid as less likely); that the code was not overlapping (i.e., that consecutive triplets did not share the same nucleotides); that the sequence of bases was read from a fixed starting point; and that the code was probably degenerate (i.e., one amino acid could be coded by more than one triplet).

However, in the end it was Marshall Nirenberg and Heinrich Matthaei who correctly figured out the correspondence between nucleotides and amino acids. They found that artificial RNAs could stimulate protein synthesis in cell-free systems. The first RNA they tried was poly-U, a long chain of nucleotides containing uracil, and they found that it contained the information for the synthesis of a protein that had many of the features of poly-phenylalanine – that is, a protein consisting of consecutive molecules of the amino acid phenylalanine. They thus concluded that one U or several Us appeared to be the code for phenylalanine. Ironically, as mentioned, the idea that a triplet consisting of the same nucleotide could code for an amino acid had been rejected by Crick and his colleagues. A few more codons were identified over the next year, and by 1965 the genetic code had mostly been resolved. The information contained in DNA and RNA sequences can be read as three-letter syllabi (triplets) called codons, each corresponding to one amino acid.

Here is how the genetic information is "expressed." DNA is transcribed to RNA, and in the case of protein synthesis this RNA is further translated into protein. Transcription of DNA to RNA is carried out by enzymes called RNA polymerases, which produce mRNA on the basis of the sequence of the transcribed DNA strand. The mRNA contains a specific sequence of nucleotides that will be translated into a sequence of amino acids. This is achieved through a sequence correspondence between mRNA and several tRNA molecules, each of which carries a specific amino acid. These molecules "meet" on ribosomes that consist of rRNA and proteins. For example, a tRNA carries the amino acid methionine and also has the triplet UAC that is complementary to the mRNA triplet AUG, which also marks the starting point for translation (Figure 2.8). Thus, in a sense, it is the tRNA molecules themselves that determine the genetic code, as they "link" the amino acids and the mRNA. However, the code is framed in terms of mRNA triplets. Among the 64 triples, UAA, UGA, and UAG do not correspond to any amino acids; translation stops there because no amino acid is added, and these are described as stop signals. This picture of the structure of the "molecular" gene and of its role in the synthesis of proteins was essentially complete by the late 1960s.

At this point, I must note that stating that DNA encodes the information for a protein does not entail that DNA "contains" all the information for the final structure of a protein. What is considered as encoded in a sequence of nucleotides is a corresponding sequence of amino acids (which is described as the primary structure of the protein). Beyond this, how an amino acid sequence folds and forms a protein also depends on the physicochemical conditions of its environment. Most proteins have a unique way of folding within normally functioning cells. However, in abnormal cells or within microorganisms in a culture dish, different three-dimensional structures are possible for the same amino acid sequence. Therefore, the three-dimensional structure of a protein (its tertiary structure) depends both on the sequence of its amino acids encoded in DNA and on the external conditions under which the folding of the protein takes place.

The "Developmental" Gene

Jacob and Monod also proposed a model for the regulation of gene expression. They made the important distinction between the "structural genes" that "determine the molecular organization of the proteins" and the "regulator and operator genes" that "control the rate of protein synthesis." According to their model of gene regulation, a molecule produced by the regulatory gene

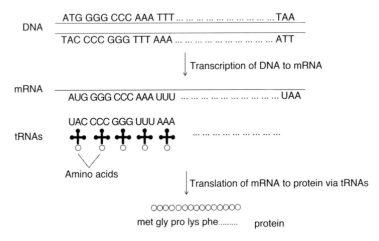

Figure 2.8 Gene expression is the flow of genetic information from DNA through RNA to proteins, or more simply the processes of the transcription of DNA to RNA and of the translation of RNA to protein. The production of RNA from DNA is described as transcription because during this process the information encoded in the sequence of the nucleotides of DNA (called deoxyribonucleotides) is copied to a similar language, that of nucleotides of RNA (called ribonucleotides). The production of proteins from RNA is called translation because the information encoded in the sequence of ribonucleotides is copied to a quite different language, that of amino acids. The rule of transcription is the specific base pairing of the nucleotides of DNA and RNA (A–T, G–C, A–U), whereas the rule of translation is the genetic code – that is, the code of correspondence between triplets of nucleotides and single amino acids, which also depends on base pairing between the nucleotides of mRNA and tRNA. In short, gene expression is based on DNA–RNA or RNA–RNA base pairing.

could block the expression of the structural genes. Thus, the production of the enzymes encoded in the structural genes did not only depend on those genes themselves, but also on other DNA sequences. The distinction between "regulatory" genes and "structural" genes was an important conceptual contribution of Jacob and Monod (they were perhaps the first to make this distinction more broadly considered, but they were not the first to think along these lines, as Barbara McClintock had made suggestions about "controlling elements" at least as early as 1951).

Another important advancement of the mid-1960s was understanding the role and nature of transcription factors, through the work of Walter Gilbert, Benno Müller-Hill, Mark Ptashne, and others. Transcription factors are proteins that bind to DNA and facilitate the binding of RNA polymerases so that transcription of DNA to RNA can begin. Thus, although all cells of an organism have the same DNA, they have different genes expressed inside them because of the different transcription factors they contain, which may also depend on the signals they receive from their immediate environment. Transcription factors already exist in the fertilized ovum, mainly derived from the mother, and are distributed to the several cells emerging from cell divisions. These proteins bind to specific DNA regulatory sequences, called promoters, which are located "before" (upstream) specific genes, eventually facilitating the binding of RNA polymerase to the promoter of a gene, and thus an RNA molecule is produced. The binding of transcription factors also depends on other proteins that bind to other regulatory DNA sequences that may be further away, called enhancers (Figure 2.9).

Finally, by 1970 several additions were being made to the "central dogma" that Crick had proposed back in 1958, through the work of David Baltimore and Howard Temin, who discovered that the synthesis of DNA from RNA was possible. Crick himself commented that these new developments, along with synthesis of RNA from RNA that had already been identified in some viruses, could be easily accommodated by the central dogma. The revised central dogma could thus describe all kinds of "transfer of sequential information." As Crick noted, the situation was not yet very clear; however, there was no reason to consider any transfer of information impossible, with the exception of that from proteins to other molecules (Figure 2.10). Thus, by the early 1970s, the central dogma could be seen as summarizing the flow of information from nucleic acids to other molecules.

Subsequent findings during the 1970s further complicated the picture. The first important finding was that changes in regulatory DNA sequences might produce changes in the respective characteristics. Thus, although the "molecular" gene that was implicated in the development of a characteristic might remain structurally unchanged, that characteristic could nevertheless change because of changes in the regulation of that gene. In other words, an organism might have a fully functional "molecular"

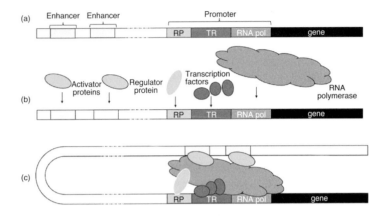

Figure 2.9   How transcription begins. Transcription factors and other proteins facilitate the binding of RNA polymerase to the promoter, and then the transcription of the gene can begin (b). This binding depends on DNA sequences both close to the gene and further away (a). The DNA molecule can actually form loops that bring activator proteins bound to enhancers closer to the site of transcription. Then the synthesis of RNA can begin (c). The dotted lines in DNA correspond to long DNA strands, which are impossible to depict in this figure.

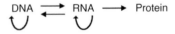

Figure 2.10   The central dogma in the early 1970s. Arrows indicate the "flow" of information on which the synthesis of new molecules is based.

gene that could, in principle, produce the protein that in turn would contribute to the development of a corresponding characteristic. But because of a mutation in the regulatory sequence, the "molecular" gene could be silenced. As a result, it would be possible for two organisms with exactly the same "molecular" gene to exhibit different versions of the same characteristic because that gene was "switched on" in one of them and "switched off" in the other.

Let's use an example to illustrate this. Imagine that there is a switch that can be used to turn the light in a room on and off. In this sense, the switch can be considered to control whether or not there will be light. Nevertheless, the light is not produced by the switch but by a bulb. Therefore, both the switch and the bulb are necessary in order to have light: The latter "produces" light and the former "regulates" its production. If there is both a functional switch and a functional bulb, then there can be light (Figure 2.11a). If either the switch or the bulb malfunctions, then there can be no light (Figure 2.11b,c). The conceptual problem here is that the "classical" gene concept could correspond either to the switch or the bulb, whereas the "molecular" gene concept can only correspond to the bulb. The reason for this is that it is not necessary to specify the nature, structure, and localization of the "classical" gene concept. As a result, a "classical" gene could be either the DNA segment that produces the respective product (the bulb) or the one that controls its production (the switch). In contrast, the nature, structure, and localization of the "molecular" gene are necessarily specified. As a result, a "molecular" gene could only be the DNA sequence that produces the respective product (the bulb). The conceptual problem, in other words, is that the "classical" and the "molecular" gene do not necessarily converge to the same DNA segment. This is possible, but it is not always the case.

Let us consider an actual example: There is an allele of the gene *Lmbr1/ LMBR1* (limb development membrane protein 1) that results in abnormal limb development in mice and humans. However, this is not the "molecular" gene that is implicated in limb development; that gene is "sonic hedgehog" (*Shh*), and it is located around 850,000 nucleotides away on the same chromosome. A sequence within *Lmbr1* called ZRS regulates the expression of *Shh* in the developing limb bud. The respective protein, Shh, acts as a signal for the patterning of the digits in a posterior to anterior direction. Therefore, if there is a change in *Lmbr1*, it is possible for an individual to have abnormal limbs without any change in the *Shh* sequence (Figure 2.12). A characteristic may thus be affected not by a mutation in the respective "molecular" gene that is directly related to it, but in some other sequence located far away that regulates the expression of the "molecular" gene. Therefore, if one thought in terms of the "classical" gene concept – a gene

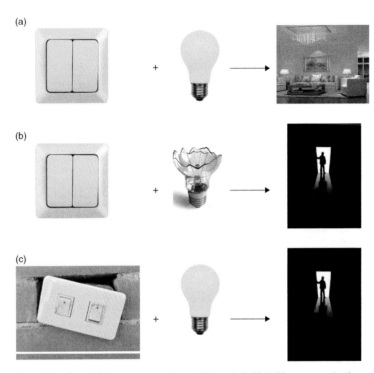

Figure 2.11   A model for gene expression and its control. Light (the gene product) requires both a functional switch and a functional bulb in order to be produced. The "molecular" gene corresponds only to the bulb that produces the light (a). In contrast, the "classical" gene can correspond either to the bulb (b) or to the switch (c). In this sense, the "classical" and the "molecular" gene do not necessarily converge to the same entity. Photographs © Steven Taylor, Oliver Cleve, Astronaut Images, WIN-Initiative/Neleman, lolostock, Achim Sass, clockwise from top left.

related to a characteristic – this could be either the *Lmbr1* or the *Shh*. However, the "molecular" gene, the expression of which is related to abnormal limb development, can only be *Shh*.

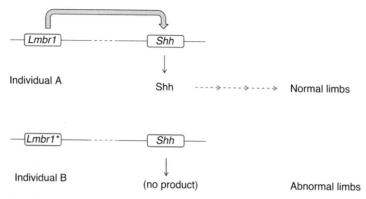

Figure 2.12   It is possible for two individuals to have the same genotype, carrying "normal" alleles of the *Shh* gene, but for one of them to have abnormal limb development due to a mutation in the regulatory sequence that is located within the *Lmbr1* gene (*Lmbr1**). The dotted line between the *Lmbr1* and the *Shh* genes indicates the long distance between them (about 850,000 nucleotides).

The finding that genes can regulate the expression of other genes brought about a new gene concept: the "developmental" gene. This can be considered as a type of molecular gene that is involved in the regulation of other genes. Developmental genes were thus considered as encoding two types of proteins: transcription factors that regulate the expression of other genes, or components of related cell-signaling processes and communication among neighboring cells. A very important implication of this conceptualization is that, in both cases, developmental genes are not independent entities but rather parts of specific pathways. This entails that the effect of such a gene is not autonomous but depends on the genes that follow along the pathway. In other words, the effect of a developmental gene actually depends on which alleles of each of the subsequent genes exist along the pathway. This means that in such cases one cannot claim that a gene G has an effect E, or that G is the gene "for" E, simply because several other genes may exist in the pathway from G to E. In this sense, different pathways are possible even if one allele is different – for example, $G \rightarrow P1 \rightarrow H2 \rightarrow A4 \rightarrow Z3 \rightarrow E1$ or $G \rightarrow P1 \rightarrow H2 \rightarrow A2 \rightarrow Z3 \rightarrow E2$ (where G, P, H, A, and Z are genes that can have different alleles, on which different effects, such as E1 and E2, are also dependent).

It should be noted that during the "molecularization" of genes, the gene concept of classical genetics did not just acquire a material, molecular identity. Neither was it altogether rejected and replaced by the molecular concept, or abandoned. In fact, the two concepts coexisted in research, depending on the interests and the explanatory aims of scientists. What was novel, though, was that thereafter genes could be conceptualized as having a concrete structure and content, as DNA molecules that carried biological information. However, soon things started to become even more complicated.

3 The Devolution of the Gene Concept

The "Exploded" Gene

During the 1970s, more puzzling observations were made. The first was that the genome of animals contained large amounts of DNA with unique sequences that should correspond to a larger number of genes than anticipated. It was also observed that the RNA molecules in the nuclei of cells were much longer than those found outside the nucleus, in the cytoplasm. These observations started making sense in 1977, when sequences of mRNA were compared to the corresponding DNA sequences. It was shown that certain sequences that existed in the DNA did not exist in the mRNA, and that therefore they must have been somehow removed. It was thus concluded that the genes encoding various proteins in eukaryotes included both coding sequences and ones that were not included in the mRNA that would reach the ribosomes for translation. These "removed" sequences were called introns, to contrast them with the ones that were expressed in translation, which were called exons. The procedure that removed the intron sequences from the initial mRNA and that left only the exon sequences in the mature mRNA was named "RNA splicing."

What exactly happens during the process of RNA splicing? Imagine that the following is the mRNA transcript of a DNA sequence that contains the information for the synthesis of a particular molecule:

abfhjdkthemvndjuthaxndjkdoublemvnshasjweuriotnxmclohelixnvml

At first sight, this sequence seems meaningless. However, if you look carefully you can spot three words that make sense. These are the exons of this gene (in bold), whereas the sequences between them are the introns.

abfhjdk**the**mvndjuthaxndjk**double**mvnshasjweuriotnxmclo**helix**nvml

This mRNA contains the specific information for the synthesis of a particular structure. The cell can "extract" this information by producing a mature mRNA molecule that will include only the sequences that correspond to the exons and will look like this:

the double helix

In this way, the information that is initially "encrypted" within a long sequence is transmitted in a clear and straightforward manner to the cytoplasm for the production of proteins. RNA splicing is performed by spliceosomes, which are ribonucleoprotein complexes that recognize exon–intron boundaries in the initial (precursor) mRNA (pre-mRNA) molecules. Spliceosomes remove the introns and assemble the exons together, thus producing the mature mRNA molecule (Figure 3.1).

As genes include sequences that correspond to amino acids but also ones that do not, whether a mutation will cause a change in the respective protein or not depends on where it occurs. A nucleotide change in an exon might affect the protein produced and the related characteristic, if such a change resulted in a different amino acid being incorporated in the protein. This is, for instance, how sickle cell anemia, discussed in the previous chapter, occurs. However, it would also be possible for a nucleotide change in an exon to produce no change in the characteristic, because both the old and the new triplet in which the change took place correspond to the same amino acid, thus resulting in no change in the protein. One might think at this point that only the exons matter as they contain coding information, whereas introns do not. However, this is not the case, as introns include sequences that have other roles. One such example has been mentioned previously: The sequence ZRS that affects the expression of *Shh* is located within intron 5 of *Lmbr1*.

What made things even more complicated was the subsequent finding that the same RNA transcript can undergo splicing in different ways, and thus

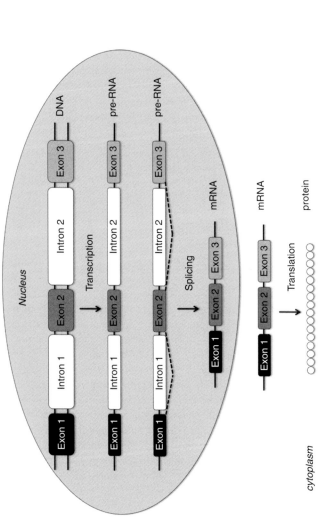

Figure 3.1 The processes of gene expression and splicing in eukaryotic organisms. DNA is transcribed to a pre-mRNA, which is spliced to produce a mature mRNA molecule, which in turn exits the nucleus and is translated to proteins in the cytoplasm (dotted lines indicate the parts of the gene that will be connected after whatever is within the edges of the dotted lines is removed).

produce different mRNAs that correspond to different proteins produced after translation. This is called alternative splicing and it results in the production of partially similar proteins from the same gene, as in the case of antibodies. Alternative splicing was initially regarded as an exception, but we currently know that more than 90 percent of protein-coding genes with many exons undergo alternative splicing. One extreme and impressive case is the *Dscam* (Down syndrome cell adhesion molecule) gene of *Drosophila melanogaster* that can produce more than 36,000 distinct mRNA molecules! Therefore, the "molecular" gene is not a linear structure that corresponds to a single RNA or protein molecule, but a modular structure that can be used in different ways to make different molecules.

Here is an illustration of alternative splicing. Imagine that the following is the sequence of the mRNA transcript of a gene:

> rhyoufnpwmslonlycndksufjdkslivencmandprldhahahletmxsjlalstwi-
> cencmdoedievnm

If you look closely you will be able to spot several meaningful words in this otherwise meaningless text:

> rh**you**fnpwmsl**only**cndksufjdks**live**ncm**and**prldhahah**let**mxsjlals**twi-
> ce**ncmdoe**die**vnm

If you are a James Bond fan, you probably recognize that the appropriate combinations of these words would produce the titles of two films: the 1967 film *You Only Live Twice* with Sean Connery and the 1973 film *Live and Let Die* with Roger Moore. With plenty of imagination, you could see this gene as the story depository of Ian Fleming, who invented the fictitious MI6 secret agent and wrote the respective stories. This depository could be expressed differently and produce distinct products (the various stories), which would have some common aspects such as the word "live" in their titles (and of course James Bond as the central character). Thus, from the mRNA stemming from this James Bond gene, splicing could produce either mRNA1 – **you only live twice** or mRNA2 – **live and let die**. It is in this sense that eukaryotic genes contain various elements, which can be combined in different ways and yield different products. Figure 3.2 presents the main types of alternative splicing.

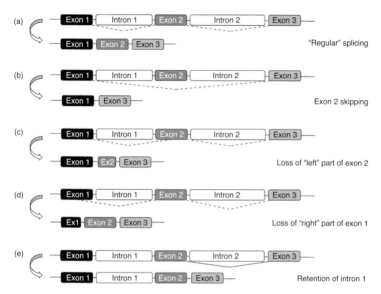

Figure 3.2   A eukaryotic mRNA (a) consisting of three exons and two introns. Splicing regularly occurs by removal of the introns, leaving only the exons in the mature mRNA (dotted lines indicate the parts of the gene that will be connected after whatever is within the edges of the dotted lines is removed). In exon skipping (b), an exon is spliced out of the transcript together with its flanking introns; this is the most common type of alternative splicing in complex eukaryotes. In alternative 3′ splice site selection (c) and alternative 5′ splice site selection (d), two or more splice sites are recognized at one end of an exon (3′ indicates the "left" part of the exon and 5′ the "right" part of the exon; "ex" in the figure indicates that only part of the respective exon was retained in the mature mRNA). Finally, in intron retention (e) an intron remains in the mature mRNA transcript. As a result, four different proteins can be produced from exactly the same "molecular" gene.

Things became even more complicated in the case of trans-splicing, a phenomenon initially observed in unicellular organisms (the "typical" splicing discussed earlier is described as cis-splicing, where cis indicates that the spliced exons come from the same RNA transcript, whereas "trans" indicates that they come from different RNA transcripts.) During trans-

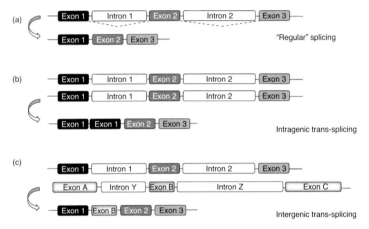

Figure 3.3   Intragenic and intergenic trans-splicing. In the first case (b), exons from two pre-mRNAs of the same gene are combined, leading to a duplication of exon 1 in the resulting mRNA. In the second case (c), exons from two pre-mRNAs from two different genes are combined, leading to an mRNA that includes exons from both genes.

splicing, exons from different mRNAs produced from different genes are joined to form a mature mRNA molecule. In other words, parts of two genes are combined to produce a single mRNA molecule. Different types of trans-splicing have been described in eukaryotes, including intragenic and intergenic trans-splicing (Figure 3.3). Intragenic trans-splicing occurs when two identical pre-mRNAs from the same gene are spliced together to generate an mRNA with duplicated exons. Intergenic trans-splicing occurs when an mRNA is produced from two pre-mRNAs derived from different genes.

Finally, even more significant complications emerged when it was found that the sequences of two "molecular" genes might overlap. In this case, the same DNA sequence can be "read" in two different ways and two distinct molecules can be produced. Initially, it was found that in a bacterial virus, mutations for a gene related to the destruction of the bacterial cell that the virus infects were located within the DNA sequence related to another protein produced in the infected cell. The interpretation given was that the two genes overlapped and that the same DNA sequence was transcribed and translated in two different ways. Several overlapping genes have been found in several other organisms, including humans. In many cases, genes are on the same DNA strand, but encode different proteins. In other cases, genes overlap partially over, for example, their first exon sequences. A special case of overlapping genes are nested genes: In this case, a gene is included within another gene. In eukaryotes, nested genes are usually located within an intron of a "host" gene. Several nested genes have been found on all human chromosomes. The exons of one gene are usually included within the introns of another gene, and the two overlapping genes are encoded on opposite DNA strands. Figure 3.4 presents some types of overlapping genes. An interesting implication here is that a single molecular change, such as a base deletion, might produce two different alleles of the two overlapping genes.

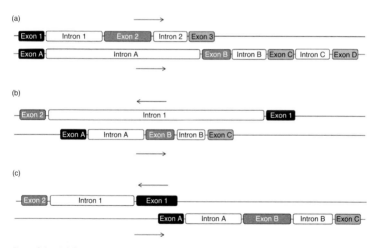

Figure 3.4   (a) Overlapping genes on the same DNA strand; (b) nested genes, where one is included in the intron of the other; (c) partially overlapping genes over their first exons. In all cases, the arrows indicate the direction of transcription. When the arrows are parallel, the genes are on the same DNA strand; when the arrows are antiparallel, the genes are on the opposite DNA strands.

These are just some of the phenomena that make implausible any talk about a one-to-one correspondence between genes and proteins. Given this, it must have now become clear why the search for "genes for" characteristics becomes pointless. Genes can at best be conceptualized as recipes that can be implemented in various ways by the cook who implements them – and in this section we have only seen a few. Only some of the ingredients may be used (splicing), the same ingredients can be used in various combinations (alternative splicing), ingredients from different recipes can be combined (trans-splicing), and some of the ingredients can overlap in the recipe so that they are not easily discerned (overlapping genes). In this sense, genes do not at all determine characteristics, in the same way that recipes do not determine the meal you will actually eat.

The findings presented in this section show why it is impossible for genes to be considered as distinct segments of DNA. Molecular biologist François Gros described this as the "exploded gene" ("le gène éclaté"). As he put it: "what best characterizes the gene today, it is not its physical and chemical materiality at the level of DNA ... it is much more *the products* that result from its activity: the cytoplasmic messenger RNA and protein." By the late 1980s, it had become clear that a gene was better characterized by what it "did" rather than by what it "was." However, this was the time that a project with a rather different perspective and a main focus on genes was about to begin.

The Human Genome Project Gene

Even though the complexities described in the previous section had already become clear in the late 1970s and the 1980s, there were still those who believed that it was possible to identify the genes in our genome and that this would be worthwhile – among them several prominent biologists and Nobel Laureates. For instance, David Baltimore wrote in 1984 that "Once the structure of DNA was found, the link between physiology and genetics was soon made. The cell's brain had been found to be a tape reader scanning an array of information encoded as a linear sequence, that is ultimately translated into three-dimensional proteins." According to Baltimore, the brain of the cell is the chromosomes that consist of DNA in which all the information is encoded: It is the "executive suite" that guides the "factory floor" of the cell. This kind of thinking paved the way for the Human Genome Project (HGP). James Watson, the first director of the HGP, noted in 1992: "Since we can now produce good genetic maps that allow us to locate culprit chromosomes and then actually find the genes for disease (as Francis Collins found the gene for cystic fibrosis), genetics should be a very high priority on the agenda of NIH research." Walter Gilbert, codeveloper of a famous DNA sequencing method, considered genes to determine more than disease:

> Three billion bases of sequence can be put on a single compact disk (CD), and one will be able to pull a CD out of one's pocket and say, "Here is a human being; it's me!" But this will be difficult for humans. To recognize that we are determined, in a certain sense, by a finite collection of information that is knowable will change our view of ourselves.

The basic premise of the HGP was that phenomena at the organismal level (e.g., characteristics) can be explained by reference to phenomena at the molecular level (e.g., genes), an idea that is described as reductionism. This idea is based on the assumption that there is a causal chain that begins from the lower levels of organizations (e.g., genes) and proceeds through the subsequent hierarchical levels (cells, tissues, organs) to produce an outcome such as a phenotype at a higher level (e.g., organism). During the 1980s, researchers such as Francis Collins, who in 1993 succeeded Watson as director of the HGP, unraveled the genetic basis of several diseases such as a fetal form of thalassemia, cystic fibrosis, and neurofibromatosis. This enhanced the view that genes cause disease. But this is misleading, as Evelyn Fox Keller has cogently noted:

> Genes are identified and located by their failure – that is, through the study of mutations, which function, if you will, as maps for misreading. We might even say that mutations "cause" such misreadings. But the further claim that the normal allele (that genetic form which is not mutant) "causes" a proper reading does not follow, either logically or physiologically. Such an inference appears to make sense only to the extent that the entire physical–chemical apparatus of the organism and its environment are effaced.

Keller has also noted that it was the concept of genetic disease, and the prospect of identifying "disease-causing" genes that made the HGP look both reasonable and desirable. Most importantly, Keller argued, the only practical application of finding a gene related to disease could be diagnostic, as treatment or prevention would require more than simply identifying a gene. However, HGP enthusiasts argued that the whole endeavor was worth the cost and effort. As James Watson noted: "We have to convince our fellow citizens somehow that there will be more advantages to knowing the human genome than to not knowing it."

The primary goal of the HGP, which started in October 1990 and was planned to last for 15 years, was to identify all genes in the human genome, which were initially estimated to number around 100,000. The first main aim of the HGP was to construct two kinds of maps: genetic and physical. Genetic maps represent the chromosomal location of genes or other identifiable fragments of DNA called DNA markers. These are constructed by figuring out how often the genes or markers of interest are inherited together (see Figure 2.4b, and the related text on linked genes and crossing over). Physical maps represent the distances between genes or DNA markers in terms of units of physical length such as base pairs (bp). A high-resolution physical map would consist of an orderly arrangement of DNA fragments, from one end of a chromosome to the other. Then comes sequencing: finding the order of bases in a DNA segment (in terms of the nucleotide bases A, T, C, and G). As Eric Lander, founding director of the Broad Institute of MIT and Harvard and one of the major figures in the HGP, wrote in 1996: "The Human Genome Project aims to produce biology's periodic table – not 100 elements, but 100,000 genes; not a rectangle reflecting electron valences, but a tree structure depicting ancestral and functional affinities among the human genes. The biological periodic table will make it possible to define unique 'signatures' for each building block."

The HGP changed the way biological research is done in fundamental ways, and resulted in new sequencing technologies. In the procedure called whole-genome sequencing, the sequences of short DNA fragments are determined and are then compared to one another. On the basis of their overlapping parts, the sequences of these fragments are estimated. The end result is a whole-genome sequence that is then compared to a previously developed reference genome, which is nevertheless incomplete as some regions are hard to sequence. It must be noted that errors are possible in these procedures, and so each base is usually sequenced about 30 times (and a lot more when the aim is to identify a disease-associated variant). An alternative approach is called exome sequencing, and in this case the aim is to find the sequence of the exons of genes. This is less costly than whole-genome sequencing, but it misses information about variants found outside protein-coding genes that might be important.

In both of these cases, it is possible to find the sequence of DNA without necessarily knowing whether variations at certain parts are important or not. In other words, DNA sequencing is just the means to produce a sequence of bases, a "text," which corresponds to the genome or the exome of an individual. Once such sequencing is completed, we can certainly try to read the "text" produced; however, this does not mean that we can always make sense of it. This, in turn, requires the ability to decode the text – that is, find words and meaning in a long sequence of letters. Here is an example to illustrate this:

Outcome of sequencing:
    abfhjdkthemvndjrxndjkdoublemvnshasjweuriotnhelixnvml
Outcome of decoding:
    abfhjdk**the**mvndjrxndjk**double**mvnshasjweuriotn**helix**nvml

In this example, the outcome of sequencing is a long sequence of letters that at first sight make no sense; in other words, the sequence does not seem to encode any message or information. Making sense of this sequence requires an additional procedure after sequencing, which is about finding which DNA segments are genes or regulatory sequences. This is the decoding of the genome, and it is a lot more difficult to do than simply finding the sequence, which is not simple anyway. It is important to note that the achievement of the HGP was only the sequencing and not the decoding of the human genome! What the HGP achieved was "writing down" the DNA sequence, not "making sense" of it. The completion of the sequence of the human genome was announced in 2000. Today, 20 years later, we are still in the course of decoding the human genome, and still far from achieving it. Important progress has of course been made, but there is still a long road ahead.

The publication of the sequence of the human genome was the end of a tough road, which provides an interesting case of competition in science. On the one hand, there were research groups that worked using public funds and that made their data publicly available. These were operating under the auspices of NIH, the Wellcome Trust, and other such institutions. On the other hand, there was Celera Genomics, founded by Greg Venter in 1998, that drew on the already publicly available data to go further in sequencing the genome, at a faster pace than the public consortium, without wanting to make their own data publicly available. Venter had already managed to develop new sequencing methods that allowed for the simultaneous sequencing of more than one gene. An intense race thus took place between the public consortium and Celera for two years, until Collins and Venter agreed in 2000 to make a joint announcement in the White House with President Clinton (Figure 3.5). At that point, it was estimated that the human genome contained 30,000–40,000 genes, not 100,000 as it was initially thought. The sequences were published in 2001, with the public consortium publishing their sequence in *Nature* and Celera publishing theirs in *Science*.

Lander wrote in 2011: "The human draft sequence reported in 2001, although a landmark, was still highly imperfect. It covered only ~90 percent of the euchromatic genome, was interrupted by ~250,000 gaps, and contained many errors in the nucleotide sequence." He also noted that the White House announcement of the draft human sequence took place in June 2000, "8 months before scientific papers had actually been written, peer-reviewed and published." Nevertheless, as shown in Figure 3.5, this announcement was filled with optimism, if not hype. The statement on the screen in the background should have been "sequencing the book of life," or even better "sequencing the human genome," and not "decoding the book of life." Decoding the human genome was a next step, and 20 years later it has not yet been achieved.

The HGP revealed the complexity of the genome and brought about a better understanding of the relation between genes and characteristics. The fact that the human genome did not contain 100,000 genes as had initially been thought, but less than half that, entailed that the relation

Figure 3.5   The announcement of the sequence of the human genome on June 26, 2000, was characterized by wishful thinking about what we would like to achieve ("decoding the book of life," as shown in the photo), rather than what was actually achieved at the time. From left to right, Craig Venter, who represented Celera Genomics, Bill Clinton, who was then the president of the United States, and Francis Collins, who represented the public consortium to sequence the human genome (© Joyce Naltchayan).

between genes and characteristics is a lot more complex than the simplistic "one gene ➔ one characteristic" one. I must note that the gene concept of the HGP was not substantially different from the "molecular" gene, in spite of the complexities revealed since the 1970s. It was also still thought at the time that protein-coding genes far outnumbered regulatory ones. The predominant model was that DNA encodes proteins, with RNAs having only an intermediate (e.g., mRNA) or auxiliary role (e.g., tRNA, rRNA) in the process of protein synthesis. As a result, for a long time DNA that does not encode proteins – about 98 percent of human DNA – was described as "junk" DNA. Even though "junk" may just mean "unused" and not "useless," "junk" DNA was considered as less important than protein-coding

DNA. This is a view of the genome as consisting of words (genes) dispersed within meaningless text (the rest of the genome). To illustrate this, first consider the following text:

nxyasjldoeplrptueijcmslswoprlcmaqwsxnmyflprtiifbvncqwieorflp-
xasnvmortipqaqkfldoaybymvlcldosjahdbdncmvmvmvnsalsksfotospq-
loeruiotnvmjcjsoapaisdahdjakflflforjgjksnxveurunlppgamqiroyxcritpgs-
wermlgotpgfsdyaxerioplsnformdorutheldunderstandingovndkylxfdrhi-
topwieorpscnyjrlsjrotjsqoelpaymfhsleuropsnxmkdopvnmsfkmvlspytuq-
wodjglptsensendahdjenrvbcnqqoepprkdalynxmshairoesnxmv-
sleirtvbcnnvmclsprshnvmckdopwisnsnajskdlviprpemvnclsodorpdshj-
lueirbdlmslprotndndjfkfvlvotjfjhsjslkflqrutipsmxbvncdjflrorqagenes-
modlpovbcncmdjcbsmxbvncmldoprfhglaskpeowxnymvjdlaprtuosw-
weqiorptsynamsireplmbkdshajyleoqashjfllcmcabyncmsdalqwuriofm-
cayxnmvlpdsueirplqweuiropsmyxnahsjldeuiwoqsnadnaslkgfkdospeor-
tusjasmahjdklsweiolfpreueioqpwxcnmdhsjalaperiwxmvnclpsdhjfru-
tioepwlqmsaqlsyxcnfjrutgkotpdlspqweuitpcpxansdjklsperuitombbxnas-
kewuqieplorrfjkdxncmyasweuiolptifncmldspaqweruivmcahgfjk-
ritlpldmcbsnncmcldpeuqabdfhrtffghdsllapweudjmvndklp

At first sight, this text seems to be totally meaningless. This is what DNA sequencing produces: a sequence of letters and nothing more. Actually, any DNA sequence is more repetitive and boring than the sequence shown, because it consists of only four letters (A, T, G, C). Now imagine that this is a DNA sequence (one of the two DNA strands), which contains the two exons of the fictitious human gene *UG* (understanding genes). If we search carefully, and we know what we are looking for, we will be able to spot the exons of this gene in this sequence, as indicated here:

nxyasjldoeplrptueijcmslswoprlcmaqwsxnmyflprtiifbvncqwieorflp-
xasnvmortipqaqkfldoaybymvlcldosjahdbdncmvmvmvnsalsksfotospq-
loeruiotnvmjcjsoapaisdahdjakflflforjgjksnxveurunlppgamqiroyx-
critpgswermlgotpgfsdyaxerioplsnformdorutheld**understanding**ovnd-
kylxfdrhitopwieorpscnyjrlsjrotjsqoelpaymfhsleuropsnxmk-
dopvnmsfkmvlspytuqwodjglptsensendahdjenrvbcnqqoepprkdalynxm-
shairoesnxmvsleirtvbcnnvmclsprshnvmckdopwisnsnajskdlvipr-

pemvnclsodorpdshjlueirbdlmslprotndndjfkfvlvotjfjhsjslkflqru-
tipsmxbvncdjflrorqa**genes**modlpovbcncmdjcbsmxbvncmldoprfhglask-
peowxnymvjdlaprtuoswweqiorptsynamsireplmbkdshajyleoqashjfllcm-
cabyncmsdalqwuriofmcayxnmvlpdsueirplqweuiropsmyxnahsjldeui-
woqsnadnaslkgfkdospeortusjasmahjdklsweiolfpreueioqpwxcnmdhsja-
laperiwxmvnclpsdhjfrutioepwlqmsaqlsyxcnfjrutgkotpdlspqweuitpcp-
xansdjklsperuitombbxnaskewuqieplorrfjkdxncmyasweuiolptifncmld-
spaqweruivmcahgfjkritlpldmcbsnncmcldpeuqabdfhrtffghdsllap-
weudjmvndklp

The phrase "understanding genes" consists of 18 characters, whereas the whole text above consists of 900 characters. Therefore, the sequence of the *UG* gene spans 2 percent of the whole sequence. But what about the rest? Is 98 percent of our DNA meaningless, as in the text above? Is it really "junk," perhaps the relic of our evolutionary history during which DNA sequences were simply accumulated? The answer is no, as I explain in the next section. The related findings have also substantially altered our conceptualization of genes.

The ENCODE Gene

The common emphasis on the importance of genes and DNA often makes people overlook the fact that the usual representations of DNA (see, for instance, Figure 2.7) misrepresent its natural state in cells. DNA is not "naked," but combined with proteins, thus forming the structures that we call chromosomes. These in turn consist of chromatin, a complex molecule that results from the chemical combination – not just the mixture – of DNA and proteins called histones. This had become clear already in the early 1970s through the work of Roger Kornberg. Chromatin consists of a repeating unit of eight histone molecules, known as H2A, H2B, H3, and H4. Two histones of each type come together to form what is called a histone octamer (i.e., a structure consisting of eight histone molecules). The DNA molecule is wrapped around this histone octamer two and a half times and the resulting structure is the nucleosome, which is the basic structure of chromatin (Figure 3.6). Therefore, DNA is the component of a larger structure – chromatin – with particular biological and physico-chemical properties.

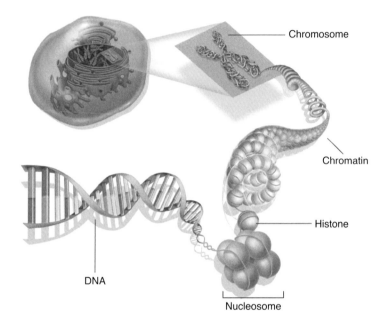

Chromosome

Chromatin

Histone

DNA

Nucleosome

Figure 3.6   The structure of chromatin and chromosomes. DNA and histones form chromatin (consisting of consecutive nucleosomes), which condenses before cell division to form chromosomes (© BSIP/UIG).

In 2003, while the HGP was being completed and it became clear that the identification of genes on the basis of the human genome sequence alone was not possible, a new project began: the Encyclopedia of DNA Elements, dubbed ENCODE. As the members of the ENCODE consortium put it: "With the complete human genome sequence now in hand . . . we face the enormous challenge of interpreting it and learning how to use that information to understand the biology of human health and disease." In other words, as I showed in the previous section, the HGP only produced a text that researchers then had to read and understand. Understanding the text in turn meant producing a "comprehensive catalog of the structural and functional components encoded in the human genome sequence." This catalog would include not only protein-coding genes, the number of which was by that time estimated to be around 20,000–25,000, but also non-protein-coding genes, transcriptional regulatory sequences, and sequences related to chromosome structure. To achieve this, the focus of the ENCODE researchers was the generation and the analysis of functional data from various experiments performed on a targeted 1 percent of the human genome.

The findings of the ENCODE project made researchers reconsider what a gene is, as the majority of the bases studied were associated with at least one primary transcript, while protein-coding DNA was just 2 percent of the bases studied. Therefore, the following definition of the gene was proposed: "The gene is a union of genomic sequences encoding a coherent set of potentially overlapping functional products." This definition summarizes three propositions: (1) that a gene is a genomic sequence (DNA, or RNA in the case of viruses) directly encoding functional molecules such as RNA or proteins; (2) that when there are several functional products sharing overlapping regions, the union of all overlapping genomic sequences coding for them should be considered; and (3) that this union is coherent – that is, done separately for final protein and RNA products, without the requirement that all products necessarily share a common subsequence. This definition is illustrated in Figure 3.7, and according to this a gene is identified on the basis of the functional products that it produces. In other words, it does not matter where the coding sequences are located, but whether their transcripts give combined functional products (either proteins or RNA).

Defining genes as DNA sequences encoding information for functional products, such as proteins or RNA molecules, means that the DNA sequence is used as the template for the production of an RNA molecule or a protein that performs some function. At first sight, the ENCODE definition of the gene seems more inclusive than older definitions, and able to cover much of the complexity of the genome. It is certainly inclusive enough to incorporate DNA sequences outside protein-coding genes, which encode non-protein-coding RNA molecules. Except for tRNA and rRNA, these include other types of RNA molecules, such as small nucleolar RNAs (snoRNAs) that are involved in RNA editing and microRNAs (miRNAs) that have important regulatory functions. But this brought about a new problem: how exactly should we

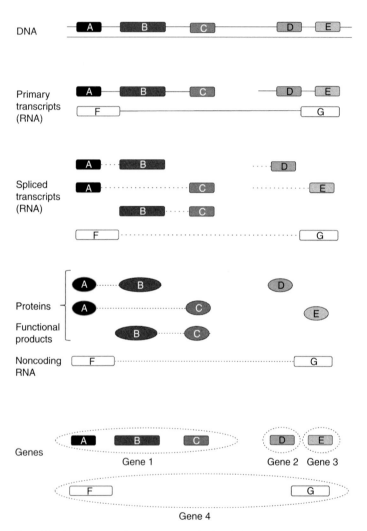

Figure 3.7 Identification of genes on the basis of their functional products. A genomic region produces three primary transcripts (ABC, DE, and FG), from which five spliced

define "function"? Under which conditions can a DNA sequence be considered to be involved in a function?

The ENCODE researchers went as far as to claim that 80 percent of the genome is transcribed and produces RNA molecules; in other words, 80 percent of the genome is functional! As stated in the abstract of the respective article in *Nature*: "The Encyclopedia of DNA Elements (ENCODE) project has systematically mapped regions of transcription, transcription factor association, chromatin structure and histone modification. These data enabled us to assign biochemical functions for 80 percent of the genome, in particular outside of the well-studied protein-coding regions." In particular, this report states that researchers integrated results from experiments with 147 different cell types and other resources, such as genome-wide association studies (see the next section). Some of their conclusions were that 80.4 percent of the human genome participates in at least one biochemical event in at least one cell type, and that there were thousands of regions that had the features of enhancers or promoters and were thus related to gene expression. Suggesting that most of the human DNA has some function, by raising the amount 40 times (from 2 percent to 80 percent), entailed that most of what was in the past perceived as "junk" DNA was not junk at all. But this was a conclusion that received harsh criticism.

One criticism focused on how the ENCODE researchers defined "function." According to the critics, the argument of the ENCODE researchers was that

---

Caption for Figure 3.7 (cont.)

mRNAs are produced (AB, AC, BC, D, and E). These are eventually translated to produce five corresponding proteins (AB, AC, BC, D, and E), whereas a noncoding RNA also exists (FG). DNA and RNA sequences are depicted as boxes, whereas proteins are depicted as ovals. On the basis of these products, four genes are identified: gene 1, consisting of segments ABC; gene 2 consisting of segment D; gene 3 consisting of segment E; and gene 4 consisting of segments FG (based on Gerstein et al.'s "What is a gene, post-ENCODE? History and updated definition," p. 678; redrawn here with the kind permission of Mark Gerstein).

DNA segments with biological functions (for instance, the regulation of transcription) tend to display certain properties (e.g., transcription factors bind to them). Therefore, if a DNA segment is found to have the same property, it can also be considered as functional. Critics argued that this argument is false because a DNA segment may display such a property without necessarily being functional, as, for example, a transcription factor may bind to a DNA sequence without necessarily resulting in transcription. Function, according to the critics, makes sense only in the light of evolution, and the functionality of a certain DNA sequence may change along its course. Thus, they proposed an evolutionary classification of genomic sequences for this purpose. They first divided the genome into functional DNA, which has a function that is the outcome of natural selection, and rubbish DNA, which does not have a function. They further divided each of these two categories into two more. Functional DNA was divided into literal DNA, in which the exact order of nucleotides is under selection, and indifferent DNA, in which only the presence or absence of the sequence is under selection. They also divided rubbish DNA into junk DNA, which does not affect the survival and reproduction of the organism and thus evolves without being selected, and garbage DNA, which affects the survival and reproduction of its carriers but for the moment exists in the genome. The important point that these authors made was that which of these categories a certain DNA sequence belongs to may change during evolution, and thus functionality should only be determined on its current status.

Another criticism made two important points, among others. The first is the importance of distinguishing between a function that is the outcome of natural selection in the recent past and that is under selection now, and a function that is considered for its current perceived role without considering past or current selection. The problem in the case of the ENCODE project, according to this criticism, is that the distinction between these two conceptions of function was not explicitly considered. A second point was that an essentialist way of thinking had been adopted by the ENCODE researchers that, according to this critic, encouraged: (1) the attribution of functions known for only a few DNA sequences to a whole class of such DNA sequences; and/or (2) the attribution of a function known for a certain part of a DNA sequence to its whole length. The problem here is that such attributions may extrapolate, without any justification, the functional significance of a certain DNA sequence, and attribute to it many more functions than it actually has.

A response to these criticisms has been that the differential expression of RNAs, including extensive alternative splicing, is a far more accurate guide to revealing the functional sequences of the human genome than assessments of the conservation, or lack of conservation, of DNA sequences through natural selection. In addition, the evidence for the biological function of noncoding RNAs in different developmental and disease contexts is – according to the defenders of this view – adequate in order to draw broader conclusions about the possible functions of other similar sequences. In this view, the criticism of the conclusions of the ENCODE researchers was not due to the lack of the relevant evidence, but other reasons. The first was the long-standing protein-centric conception of gene structure and regulation in human development, under which research has been conducted since the 1940s and which should be reconsidered. The second reason was that the idea of a largely nonfunctional genome full of "junk" DNA has been used as evidence for the evolutionary accumulation of DNA sequences and against any notion of intelligent design. Such an argument is obviously threatened if one accepts that 80 percent of the human genome is functional.

The traditional way to assess functionality has been to compare the DNA sequences of different species and document the changes. In the case of humans, functionality can be inferred by comparing human DNA sequences with the respective ones in model organisms. The main assumption of this approach is that functional elements should not be very different between different species. The reason for this is that changes may affect functionality in detrimental ways; therefore, it is unlikely for such changes to have been conserved in the course of evolution. In contrast, changes in nonfunctional parts of the genome may have no impact at all, and therefore may have been accumulated in a lineage in the course of evolution. If one applies these criteria, one finds conservation of protein-coding sequences (up to 85 percent between humans and mice), but little conservation outside them. Overall, this suggests that only 3 percent of the whole genome is functional, but this might overlook species-specific functional sequences. Another way to assess functionality is by measuring biochemical activity. This was the approach taken by the ENCODE consortium, following which they concluded that 80 percent of the genome is functional. Obviously, it is not simple to reconcile the two approaches and decide what is the case.

Beyond this debate, there are two important conclusions from the ENCODE project and the relevant research that are important. First, there exist RNAs that do not code for proteins and that have very important roles in cells. In late July 2020, the ENCODE researchers announced that they had identified 20,225 protein-coding and 37,595 noncoding genes in the human genome. Whereas transcription factors regulate gene expression only at the transcription level, noncoding RNAs affect this process at multiple levels. Second, and perhaps most importantly, it became clearer that it is far from simple to define and identify genes. Whether or not one agrees with the assumptions of the ENCODE project, what is important is that the information encoded in human DNA related to non-protein-coding/regulatory sequences is at least as important as the respective information for proteins. Therefore, there are multiple ways in which DNA may be related to our characteristics. As the relation between genes and gene products has been found to be complex, researchers in genomics focused anew on changes in DNA sequences across the whole genome and their impact.

Genome-Wide Association Studies and the "Associated" Gene

As already mentioned, the HGP was motivated by the premise that finding the sequence of human DNA might provide important clues for understanding disease. During the last 15 years or so, there has been a lot of interest in studies of distinct groups of people that differ in one characteristic, in which researchers look for whether particular genetic variants are more common in one or the other group. These studies are called genome-wide association studies (GWAS). Their focus is often on variations down to the level of a single nucleotide, which are called single nucleotide variants (SNVs). We have already seen in Chapter 2 how point mutations – that is, changes of individual bases in the DNA sequence – are possible. When the changes in the DNA sequences do not bring about any significant change in a gene product, they are not described as mutations; rather, they are described as SNVs, simply because they vary between individuals in a population. When an SNV is found to exist at a frequency higher than 1 percent, it is described as a single nucleotide polymorphism (SNP).

It should be noted that SNPs are not the only type of variants that can be found in DNA sequences. There also exist insertions–deletions, block substitutions, inversions of DNA sequences, and copy number differences (Figure 3.8), which can be identified with various molecular methods. However, these constitute only a small portion (0.5–1 percent) of any given genome, and therefore SNPs are the most prevalent type of genetic variation among individuals. Although rare and novel SNPs in individuals also exist, when the

Figure 3.8 DNA variants in the genome. Single nucleotide variants are DNA sequence variations in which a single nucleotide (A, T, G, or C) is altered. Insertion–deletion variants occur when one or more bases are present in some genomes but absent in others. Block substitutions are cases in which several consecutive nucleotides vary between two genomes. An inversion variant is one in which the order of the base pairs is reversed. Finally, copy number variants occur when identical sequences are repeated and the difference is thus in how many times the repeated sequence exists in the genome.

genomes of two individuals are compared, the majority of the nucleotides that differ are at positions of variants common in the whole population. The full sequencing of human genomes has shown that in any given individual there are, on average, four million genetic variants. The important task, of course, should be to find out which of these variants actually affect phenotypes, and not just which ones are associated with them.

Genome-wide association studies use dense maps of SNPs that cover the human genome in order to look for differences in the frequencies between different groups of individuals. If a genetic variant is more common in one group, and this difference is a statistically significant one, then an association is said to have been found between that genetic variant and the characteristic of that group. It is also assumed that a significant difference in the frequency of a variant between two groups indicates that the corresponding region of the genome contains functional DNA sequences that somehow affect the respective characteristic. The main principle behind GWAS is linkage disequilibrium, the nonrandom association of alleles at two or more loci (Figure 3.9). I must note that in this case "allele" means something different than what it meant in the eras of classical and molecular genetics. The alleles are no longer found at the level of the gene, but far below it in short sequences and even single SNPs.

So, what does nonrandom association of alleles mean? The idea is that a genetic variant, which somehow affects a characteristic and which is shared by some individuals because of common descent, will be surrounded by other shared genetic variants in a sequence that should correspond to that in the ancestral chromosome on which they first occurred. In other words, certain alleles are found close to one another not by chance but because they once were, and still are, linked, and thus tend to be inherited together (these constitute a haplotype). The reason for this is that recombination through crossing over can rearrange alleles up to a certain point; some genetic variants are so close that they cannot be separated by crossing over. As a result, when researchers are looking for genes that are potentially associated with a characteristic, they can first look for SNPs associated with it. When such SNPs are found, it is possible that a genetic variant somehow involved in the production of the characteristic is located close to them.

It must be noted that once the characteristic of interest is associated with one or more SNPs, these are studied in independent samples for statistical validation. This is important because when thousands of SNPs are tested, many of them could be found to be statistically significant just by chance. For instance, if a study includes 500,000 SNPs, it is expected that 25,000 of them could be found to be statistically significant due to chance for a significance threshold of $p < 0.05$. It must also be noted that in a GWAS the lower the frequency of the less common variant, the lower the ability to detect an association between that and the condition. This is why GWAS has detected associations with relatively more common variants. In short, GWAS can point to associations that do not really exist, and cannot identify associations between characteristics and rare variants. These are important limitations worth keeping in mind. One way to address these problems has been to adopt a standardized significance threshold of $p < 5 \times 10^{-8}$. This means that in most current studies the probability that an identified association has occurred by chance is 5×10^{-8}. To return to the example above, for a study that includes 500,000 SNPs, it is not expected to find any associations by chance with such a threshold.

The findings of GWAS are easy to misunderstand if one fails to recognize a crucial distinction: It is one thing to account for the variation of a characteristic on the basis of particular SNPs, and another to causally explain how those DNA sequences affect the characteristic. This means that finding associations between SNPs and characteristics may be informative, but does not reveal much about the underlying biological mechanism – that is, why and how the SNP affects this characteristic. In short, association does not entail causation. Therefore, whenever researchers announce that they have found associations between an allele or SNP and a particular characteristic, it means that they have only found a statistical association between them. This does not necessarily entail that the allele or SNP somehow causes the characteristic. The reason is that a statistical association between two variables may exist without a cause–effect relation between them.

Let me use a simple example to explain this. The sales of 0.5-liter bottles of cold water and of ice creams rise significantly in Greece during summer. Therefore, one might find a statistical association between the number of ice creams and bottles of cold water sold during summer compared to, say, the

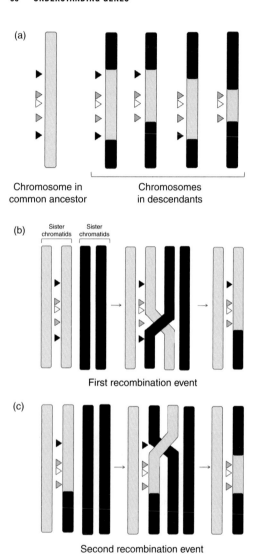

(a)

Chromosome in
common ancestor

Chromosomes
in descendants

(b)

Sister
chromatids

Sister
chromatids

First recombination event

(c)

Second recombination event

Figure 3.9   (a) Linkage disequilibrium around a genetic variant, the position of which is indicated by the white triangle. This existed on the chromosome of a common ancestor,

respective sales during winter. Can there be a cause–effect relation there? The answer is yes and no; or better: it depends. It is possible that there is a real cause–effect relation if most people who eat ice cream also need to drink cold water after doing that. In this case, one can assume that the increase in the consumption of ice cream has a causal influence on the increase in the consumption of cold water, because it is the consumption of the former that causes the consumption of the latter. However, it could also be the case that both consumptions increase independently because of the high temperatures in Greece during summer (often higher than 35 °C), and thus because people need to have something chilled to freshen up. In this case, there is no cause–effect relationship between the consumption of ice cream and cold water. The causal factor is the high temperature that makes more people buy either or both of these. Therefore, in order to understand what is going on, we need to study in detail the underlying process. Returning to genes, finding an association between gene A and condition C and between gene B and condition C is not really informative if we do not know what exactly A and B contribute to C. Do A and B make independent contributions to C? Or is the contribution of B dependent on the contribution of A, because A affects B? Or

---

Caption for Figure 3.9 (cont.)

and was linked to some other genetic variants, the positions of which are indicated by the black and gray triangles. The ancestral chromosome in the common ancestor included all these variants, whereas the chromosomes of the descendants do not include all of them because some have been moved away due to consecutive recombination events, presented in (b) and (c) (the part of each chromosome that is similar to the ancestral one is shown in gray, whereas the new parts introduced due to recombination events are shown in black). The genetic variants indicated by the black triangles may have been removed from some chromosomes because of recombination. In contrast, those variants indicated by the gray triangles are physically closer to the "white" genetic variant and so remain associated with it, despite the modifications brought about by recombination events. Therefore, these variants will be nonrandomly associated with the "white" genetic variant because they are linked. This property is used in GWAS as researchers can look for genetic variants and see if these are associated with a particular characteristic. Once these are found, it is possible to also locate the gene associated with the characteristic. Figure (a) is adapted by permission from Macmillan Publishers Ltd: *Nature Reviews Genetics* (Kruglyak, 2008), © 2008.

do both *A* and *B* affect an unknown gene *U*, which is actually the major contributor to condition C? Until the precise biological process is understood, associations are not very informative on their own, because different biological processes may underlie the same association (Figure 3.10).

This, in turn, requires the identification of the respective causal factors: what caused a certain characteristic or a disease to develop. Genes can certainly have a causal role in such cases, but they are not the only causal factors. There can be several kinds of causal factors, but which ones are selected from

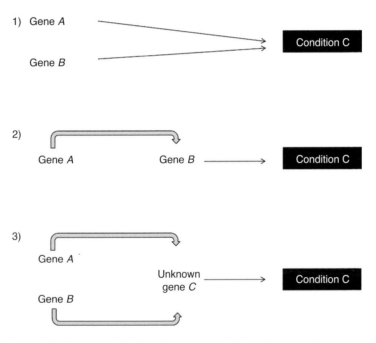

Figure 3.10   Some possible biological processes underlying a documented association between genes *A* and *B* and an observed condition C. Until one knows which of the processes (1), (2), (3) is the correct one, there is no real understanding of the effect of genes *A* and *B* on condition C.

the set of factors sufficient to produce an outcome depends on the contexts and on our interests. Let us consider a broken window as an example. If a door with eight glass panes slams and one of the panes breaks, there are several different causal questions that might be asked. If the question asked was why the glass pane broke, then the slamming of the door can be a sufficient causal explanation. But the question asked could be why it was the particular pane that broke and not any of the other seven. In this case, there are several potential explanations. It could be that this pane was more fragile because it was made of a different type of glass from the other seven panes; it could be its location on the door that made it more prone to breaking; or it could be that the framing around that pane was not that good. Finally, we might be interested in why the door slammed. This could happen because someone opened the window in the room and there was a gust of wind, or because the people living in the house had a fight and one of them slammed the door. In both cases, the causal inquiry can go further, to a physical explanation of what differences in the temperatures of aerial masses caused the wind, or a psychological explanation of why there was a fight between the people. Our world is a multicausal one.

Notwithstanding the original expectations of GWAS, their findings have produced two important conclusions:

- *Many genes* ➔ *one characteristic*: The so-called "complex" characteristics are highly polygenic – that is, are influenced by several genes. At least 10,000 statistically significant associations (at the genome-wide p-value threshold of 5×10^{-8}) between genetic variants and one or more complex characteristics have been found. The conclusions from GWAS has been that for most complex characteristics that have been studied, variants in many genes contribute to genetic variation in the population.
- *One gene* ➔ *many characteristics:* The fact that most characteristics are associated with multiple genetic variants entails that several of these variants are the same and therefore affect several characteristics. This has been found in different kinds of studies with different kinds of variants. The true nature of this phenomenon, called pleiotropy, is currently unknown but, at least in some cases, could imply an impact of the variants on different tissues and/or at different ages.

In Chapter 4, I provide concrete examples as evidence that many genes affect the same characteristic, which clearly shows that there are no "genes for" characteristics or disease.

In recent years there has been a gradual move from a gene-based epistemology to a sequence-based one. But if we insist on using gene concepts, in my view research in genomics has brought about a new gene concept that I call the "associated" gene. The "associated" gene is a DNA sequence, which can be as small as an SNP and which has been found to be associated with, and sometimes causally related to, a characteristic or a disease. Whence this concept? On the one hand, scientists often describe the various SNPs as alleles; on the other hand, genetic tests are supposed to assess the impact of these SNPs on our lives by considering their contributions to the probability of developing a disease. The "associated" gene actually has properties of both the "classical" and the "molecular" gene. Like the "classical" gene, it accounts for differences in phenotypes; and like the "molecular" gene, it can be related to phenomena and processes at the molecular level insofar as the underlying biological mechanism is understood. At the same time, it is very different from both the "classical" and the "molecular" genes because contrary to those it has a statistical dimension, due to the associations between SNPs and characteristics revealed by the GWAS.

Actually, the "associated" gene may not be a gene at all. But it serves the same role that the "classical" and the "molecular gene" concepts used to serve, especially with regard to characteristics and disease. Whether or not this concept is the most important for genomics research, it seems to be the most relevant for the public understanding of genes. Therefore, it is a concept that nonexperts need to understand, and to become aware of both its value and its limitations. This is absolutely necessary in order to refrain from shifting from genetic fatalism perceptions (essentialism, determinism, and reductionism) related to single genes "for," to genomic fatalism perceptions related to multiple SNPs spread within the genome.

The ENCODE project and GWAS paved the way for the postgenomic era, or simply postgenomics. As historians Sarah Richardson and Hallam Stevens have noted, "For many, postgenomics signals a break from the gene-centrism and genetic reductionism of the genomic age." The gene of the postgenomic era is no longer considered as a distinct segment of DNA, but something more inclusive: "A postgenomic gene is the collection of sequence elements that is the 'image' of the target molecule (the product whose activity we wish to understand) in the DNA, however fragmented or distorted this image may be." The findings of research in genomics have also brought about a significant transformation of our conceptualization of the genome. Evelyn Fox Keller has suggested that this has been the shift from the pregenomic genome, conceived as a static collection of active genes – separated by "junk" DNA – initiating causal chains leading to the formation of characteristics, to the postgenomic genome, a dynamic and reactive system dedicated to the regulation of protein-coding sequences of DNA on the basis of changing signals received from the environment. Overall, the HGP, the ENCODE project, and GWAS have shown that we should look at the genome as a whole, and not at individual genes, to understand hereditary phenomena.

4   There Are No "Genes For" Characteristics or Disease

Biological Characteristics: Eye Color and Height

If you were taught Mendelian genetics at school (see Figures 2.1 and 2.2) you should be aware that it is an oversimplified model that does not work for most cases of inherited characteristics. Human eye color is a textbook example of a monogenic characteristic. It refers to the color of the iris – the colored circle in the middle of the eye. The iris comprises two tissue layers, an inner one called the iris pigment epithelium and an outer one called the anterior iridial stroma. It is the density and cellular composition of the latter that mostly affects the color of the iris. The melanocyte cells of the anterior iridial stroma store melanin in organelles called melanosomes. White light entering the iris can absorb or reflect a spectrum of wavelengths, giving rise to the three common iris colors (blue, green–hazel, and brown) and their variations. Blue eyes contain minimal pigment levels and melanosome numbers; green–hazel eyes have moderate pigment levels and melanosome numbers; and brown eyes are the result of high melanin levels and melanosome numbers. Textbook accounts often explain that a dominant allele B is responsible for brown color, whereas a recessive allele b is responsible for blue color (Figure 4.1). According to such accounts, parents with brown eyes can have children with blue eyes, but it is not possible for parents with blue eyes to have children with brown eyes. This pattern of inheritance was first described at the beginning of the twentieth century and it is still taught in schools, although it became almost immediately evident that there were

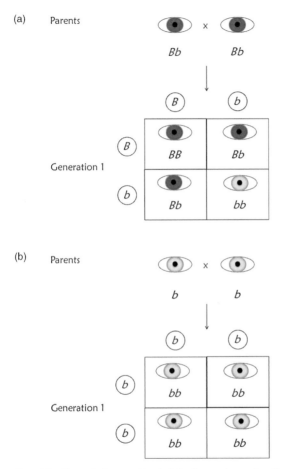

Figure 4.1   The typical account for the inheritance of eye color often found in textbooks. According to this account, a dominant allele determines brown color and a recessive allele determines blue color. Thus, parents with brown eyes can have children with brown or blue eyes (a), whereas parents with blue eyes can only have children with blue eyes (b) (brown eyes are shown in this figure as darker than blue eyes). However, this model of inheritance is insufficient to explain the observed variation.

exceptions, such as that two parents with blue eyes could have offspring with brown or dark hazel eyes.

However, the inheritance of eye color, actually of iris pigmentation, is not that simple. More than one gene has been found to be significantly associated with eye color. The strongest associations were initially found between eye color and the *OCA2* gene (OCA2 melanosomal transmembrane protein, involved in the most common form of human oculocutaneous albinism, a disease characterized by fair pigmentation and susceptibility to skin cancer). The *OCA2* gene is located on chromosome 15 and encodes a protein affecting melanosome maturation. Other strong associations between eye color and particular genes have also been found, and so eye color is best described as a polygenic characteristic – that is, one to which multiple genes contribute. Nevertheless, it seems that particular genetic variants, closely located to one another, seem to account for blue eye color. It has been found that three single nucleotide polymorphisms (SNPs) within intron 1 of the *OCA2* gene have the highest statistical association with blue eye color. Other studies have shown that SNPs in the introns of gene *HERC2* (the official name of which is "HECT and RLD domain containing E3 ubiquitin protein ligase 2"), also on chromosome 15, are strongly associated with blue eye color. It has been assumed that the variants within the *HERC2* gene are related to the expression of *OCA2*, and that it is the decreased expression of the latter in iris melanocytes that is the cause of blue eye color. In this sense, one could say that there exist alleles "for" blue color, but these are not the alleles of a single gene. Furthermore, even if such a model worked in many cases, it would still be insufficient to explain the full range of eye colors.

Let us consider a simple example: Assume that *O* is the allele that produces the OCA2 protein and *N* is an allele that does not; similarly, let us assume that *T* is the *HERC2* allele that enhances the expression of the *OCA2* gene, whereas the *C* allele does not. We can thus assume that a woman with genotype *NNTC* and a man with genotype *ONCC* will have blue eyes because they do not produce the OCA2 protein, for different reasons. The man with genotype *ONCC* has the necessary *OCA2* allele (O), but it is not expressed because he also lacks the *T* alleles; the woman with genotype *NNTC* does not have the necessary *OCA2* allele. If these two persons decided to have offspring, what eye color would you expect those to have? The father

Parents

Generation 1

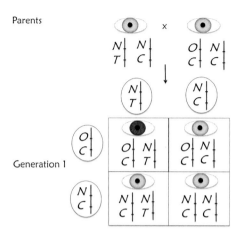

Figure 4.2   How two blue-eyed parents can have a child with brown eyes (brown eyes are shown in this figure as darker than blue eyes). In this case, the alleles of the two genes are located on the same chromosome and so are inherited together. Different combinations of the *OCA2* and *HERC2* alleles have been found in different human populations.

could produce different kinds of spermatozoa with either the *OC* or the *NC* alleles, whereas the mother could produce either an ovum with the *NT* alleles or an ovum with the *NC* alleles. The possible combinations are shown in Figure 4.2. It becomes evident that there is a 75 percent probability that their child would have blue eyes as well, if the child had any of these combinations of alleles: *NNTC* (no *OCA2* allele), *NNCC* (no *OCA2* allele), *ONCC* (*OCA2* allele exists but is not expressed). However, there also would be a 25 percent probability that the child would have the alleles *ONTC*, and carry both the *OCA2* and the *HERC2* alleles that result in brown eye color. Two blue-eyed parents would thus have a child with brown eyes.

Let us now consider another characteristic: human height. We now know from various studies that approximately 80 percent of the variation in height among individuals within a population is due to variation in genetic factors (this is called heritability and it is discussed in the next chapter). Several variants have been found to influence this characteristic. In 2008, three genome-wide association studies (GWAS) in a combined sample size of approximately 63,000 individuals concluded that 54 variants were found to be associated with height. One study used GWAS data from 13,665 individuals and identified 20 SNPs associated with adult height. However, the 20 SNPs combined explained only 3 percent of the variation in height within the population. Another study used GWAS data from 15,821 individuals and found 12 loci strongly associated with height. These 12 loci combined accounted for approximately 2 percent of the population variation in height. A third study used GWAS data from 33,992 people, and identified 27 sites of the genome with one or more variants that showed significant association with height. These variants explained approximately 3.7 percent of the population variation in height. Variants related to four genes were found in all three studies: *HMGA2* (high-mobility group AT-hook 2) on chromosome 12, producing a protein that belongs to a family of proteins that function as chromatin architectural factors; *ZBTB38* (zinc finger and BTB domain containing 38) on chromosome 3 that encodes a transcription factor; *HHIP* (hedgehog interacting protein) on chromosome 4 that encodes a protein involved in signaling during developmental processes; and *CDK6* (cyclin-dependent kinase 6) on chromosome 7 that is involved in the control of the cell cycle. The variants identified are related to a variety of pathways, including signaling, extracellular matrix, chromatin remodeling, and cancer, indicating the complexity of the biological regulation of human height.

That so many variants explained so little variation in height (2–3.7 percent in the studies cited) might indicate that other, not-yet-identified SNPs with small effects might affect this characteristic. It would also be possible that there exist variants that the GWAS approach does not capture. More recent studies have provided a more detailed picture. Whereas it is possible that unidentified genetic variants could account for the variation in height, it could also be the case that the actual effect of the identified ones was missed. Studies such as those just described analyze GWAS data by testing each SNP individually for an association with a characteristic. The statistical tests used reduce the occurrence of false associations, but may also miss real associations if the respective SNPs have a small effect. A different approach is to estimate the variance (a parameter of a distribution – the square of the standard deviation – that is mathematically useful for describing the shape of the distribution and the data spread) explained by all SNPs together. This was attempted in a study with GWAS data on 294,831 SNPs from 3,925 unrelated individuals. It was found that 45 percent of variance for height could be explained by considering all SNPs simultaneously. This was further confirmed by another study using GWAS data on 586,898 SNPs from 11,586 unrelated individuals. This study also showed that the variance explained by each chromosome is proportional to its length, and that SNPs in or near genes explain more variation than those between genes.

Obviously, numerous genetic variants contribute to human height, and so these might be identified through GWAS with larger samples. In a study that used data from 183,727 individuals, it was shown that hundreds of genetic variants, in at least 180 loci, were associated with variation in human height. These 180 loci were found to be related to 21 genes mutated in human syndromes characterized by abnormal skeletal growth. These data could explain approximately 10 percent of the phenotypic variation in height. A more recent study used GWAS data from 253,288 individuals and identified 697 SNPs in 423 loci that explained 16 percent of the observed variation. The main conclusion was that human height is influenced by a very large number of variants, found throughout the genome and related to pathways relevant to skeletal growth. More recently, a study with about 700,000 individuals, looking at approximately 3,290 near-independent SNPs associated with height, managed to explain about 24.6 percent of the variance in height.

In this section, I considered human eye color and height to show that several genes – in some cases a few, in other cases many – are implicated in their development. Therefore, the simple Mendelian model of inheritance that you learned at school does not really work for most characteristics. There are no "genes for" characteristics, but there generally exist several (a few or many) genes or other DNA sequences that somehow affect characteristics and that can be associated with their variation.

Behavioral Characteristics: Aggression

How about behavioral characteristics? Are genes associated with certain behaviors? If yes, to what extent? We should remember that we do not inherit only our DNA from our parents. As everyone knows from experience, parents (biological or adoptive) usually prescribe the immediate social environment in which a person is brought up. Thus, children actually inherit from their parents this social environment, as well as several other relevant features, including values, principles, and worldviews. Eventually, the behavior of children is shaped in this context, influenced by the behavior of their parents, as well as by several factors that are directly related to this environment, such as schooling, friends, the wider family, and more. However, it is the reference to genes that really seems to excite people when it comes to human behavior.

In 1993, a study reported findings in a Dutch family in which certain males exhibited intellectual disability and aggressive behavior that was usually triggered by anger. This behavior could last for a few days, during which these males slept very little and experienced frequent night terrors. This condition was found to be associated with a locus on the X chromosome, which seemed to be related to monoamine oxidase A (MAOA), an enzyme involved in the metabolism of neurotransmitters, such as dopamine, norepinephrine, and serotonin. Biochemical analysis indicated a disturbance of monoamine metabolism in the affected males. It was therefore concluded that a mutation affecting the *MAOA* gene might be responsible for the observed behavior. In another study published in the same year, the researchers described this mutation in more detail. In each of the five affected males tested, a single nucleotide change was identified in the eighth exon of the *MAOA* gene, which resulted in the replacement of amino acid glutamine by a stop codon, and thus to a shorter MAOA enzyme. All affected males had a C to T substitution (i.e., C had been replaced by T) at a certain position, whereas 12 normal males of the same family had C at that position.

This finding attracted the attention of other researchers who attempted to study the possible connection between the *MAOA* gene and aggressive behavior. It was already known that a polymorphism in the promoter region of the *MAOA* gene affected its expression. The polymorphism consists of a repeat (repeated sequence) of 30 nucleotide pairs, present in 3, 3.5, 4, or 5 copies. The variants with 3.5 or 4 copies of the repeat were found to be transcribed up to 10 times more efficiently than those with 3 or 5 copies. The variants with the 3 and the 4 copies were also the most commonly found ones, accounting for more than 95 percent of the observed variation. Thus, researchers decided to look at these variants of the *MAOA* promoter, the corresponding MAOA expression, and possible associations with different kinds of exposure to childhood maltreatment. In particular, researchers studied a cohort of approximately 1,000 children, about half of which were male, and which were assessed at various ages until the age of 26 years. For this sample it was known that between the ages of 3 and 11 years, 8 percent of the children had experienced "severe" maltreatment, 28 percent had experienced "probable" maltreatment, and 64 percent had experienced no maltreatment. On the basis of the hypothesis that the *MAOA* genotype can moderate the influence of childhood maltreatment on neural systems implicated in antisocial behavior, the researchers tested possible associations between variants in the promoter of the *MAOA* gene that conferred low or high MAOA expression, maltreatment, and antisocial behavior. It was concluded that severely maltreated children with the promoter variant conferring high levels of MAOA expression were less likely to develop antisocial behavior than severely maltreated children with the promoter variant conferring low levels of MAOA expression (Figure 4.3). The researchers noted that 85 percent of the latter group had developed some form of antisocial behavior, implying that this was not a coincidence. Yet, they also noted that this study only provided preliminary findings and that further studies were required to see whether these results could be replicated and confirmed.

Several studies attempted to replicate these results. A recent meta-analysis of 27 published studies, from the original 2002 one until 2012, investigated the interaction of the *MAOA* genotype and childhood maltreatment in antisocial behavior. The meta-analysis concluded that across 20 male cohorts, early adversity was associated with antisocial outcomes more strongly for low MAOA activity individuals compared to high activity ones. The authors concluded that maltreatment presumably interacts with MAOA to augment aggressive or antisocial potential, previously admitting that the biological mechanisms underlying the correlations under discussion remained unknown. Another meta-analysis of data for MAOA from 31 studies found a modest, positive association between the low activity MAOA variant and antisocial behavior.

In contrast, a meta-analysis of data from 17 studies found no significant association between the *MAOA* gene promoter variants and aggression. The authors noted, interestingly, that this lack of association could be explained by the following considerations: (1) that a complex behavior like

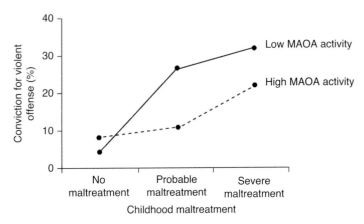

Figure 4.3   The association between childhood maltreatment and subsequent anti-social behavior as a function of MAOA activity, in the 2002 study by Caspi et al. The graph presents the percentage of males convicted of a violent crime by age 26 and childhood maltreatment in high and low MAOA activity individuals. The effect of maltreatment was found to be significant in the low MAOA activity group (163 indi-viduals), and not significant in the high MAOA group (279 individuals). Therefore, the *MAOA* gene promoter genotype seemed to affect aggressive behavior (data from Caspi et al.'s "Role of genotype in the cycle of violence in maltreated children," p. 852, figure 2b; the graph here is drawn with approximation).

aggression is explained by hundreds or thousands of genes with complex interactions, rather than a few candidate genes; (2) that aggression and violence are complex behaviors that exist on a continuum, and so it is more likely that they are determined by many genes of moderate or small effect; and (3) that the sample sizes used in the reviewed studies were small, whereas GWAS studies show that very large samples are needed to reveal interesting findings in most human characteristics and diseases. It must be noted that not only the *MAOA* variants but also others such as 5HTTLPR, the serotonin-transporter-linked polymorphic region of the serotonin transporter gene, have been considered to be related to aggression. More broadly, most candidate gene association studies have identified associations between aggressive behavior and genes involved in neurotransmission and in hormone regulation, whereas GWAS have not yet identified genome-wide significant associations.

So, what does all this mean? It is clear that the *MAOA* gene is not the "gene for" aggression; several genes might be related to this complex behavioral characteristic, and further studies are required to clarify the underlying biological processes. One important question is how exactly can we measure aggressive behavior? As this can be of different forms, such as increased anger, decreased fear, high emotional reactivity, or decreased control, "aggression" in general is perhaps a vague description of a behavior. But even if it were the *MAOA* gene promoter variants alone that affected aggressive behavior, there would still be problems in deciding how to make good use of this information. Consider again Figure 4.3. The level of convictions for violent offenses is less than 10 percent, both for the low MAOA activity and for the high MAOA activity individuals when they have not experienced childhood maltreatment. In contrast, the amount of convictions approximately triples for either of these groups when individuals have experienced severe childhood maltreatment. Therefore, doesn't childhood maltreatment look like a more important factor for aggressive behavior than the *MAOA* alleles?

Monogenic Diseases: Thalassemias and Familial Hypercholesterolemia

What is a disease? Disease cannot be defined in absolute terms, but only in relation to some state that we have agreed to consider as "normal." If we have agreed on some specific "normal" levels of hemoglobin – the molecule that transfers oxygen and carbon dioxide in blood – then lower levels resulting in insufficient transfer of these gases inside the body can be considered as a disease (such as β-thalassemia). Similarly, if we have agreed on some specific "normal" levels of cholesterol in blood, then higher levels that might in the long term be related to heart problems can be considered as a disease (such as familial hypercholesterolemia). But in both cases, these diseases emerge as deviations from what we consider as the "normal" state. Therefore, a disease is a difference in a characteristic. A factor considered to have caused a disease is one responsible for the transition from the "normal" state to the "disease" state. In both β-thalassemia and familial hypercholesterolemia, the factor responsible is assumed to be a change in a particular gene. Beyond that, the approach to understanding the development of a disease is the same as the approach to understanding the development of a characteristic: We usually first find associations between particular DNA sequences and the occurrence of particular diseases, and then we try to figure out the pathway through which these DNA sequences may bring about the respective diseases. In this section, I consider in some detail β-thalassemia and familial hypercholesterolemia, which are usually cited as examples of a "recessive" and a "dominant" Mendelian disease, respectively.

Hemoglobins are complex proteins contained in human red blood cells. All hemoglobins consist of heme, a molecule that carries oxygen, and four protein chains called globins – hence the name hemoglobin. In all cases, α-globin chains combine with other globin chains to give hemoglobins HbA ($\alpha2\beta2$) and HbA2 ($\alpha2\delta2$), which are the two main types in adults, as well as fetal hemoglobin HbF ($\alpha2\gamma2$), which also exists in adults in very small amounts. The β-thalassemias occur when there is decreased or no production of β-globin chains and thus of HbA. This is due to mutations within, or related to the expression of, the *HBB* (hemoglobin subunit beta) gene on chromosome 11 that encodes the β-globin chains. Some mutations completely inactivate the *HBB* gene, resulting in no β-globin production. Other mutations simply cause a reduction in the amount of β-globin produced. This leads to an excess of α-globin chains, which aggregate in the precursor red blood cells, causing their abnormal development, and in mature red blood cells causing membrane damage and cell destruction, which in turn results in anemia that retards growth and development. Generally speaking, the disease has been considered to be a monogenic one, exhibiting a pattern of typical Mendelian inheritance, as shown in Figure 4.4a.

Thalassemias are some of the most common genetic diseases worldwide. Thalassemia phenotypes have an extremely high clinical heterogeneity. Generally speaking, there are three main clinical phenotypes of β-thalassemia: major, intermedia, and minor. On the one hand, many homozygotes and certain compound heterozygotes (i.e., people carrying two different alleles that are both associated with the disease) have β-thalassemia major and severe anemia from early in life, which may lead to death during the first year. Only if the symptoms are controlled by blood transfusion that provides the missing hemoglobin, and only if the excessive amounts of iron thus administered are removed, can children survive to adulthood. On the other hand, heterozygotes (i.e., those carrying one β-thalassemia allele) and other compound heterozygotes can have a variety of conditions ranging from β-thalassemia major and severe anemia to mild β-thalassemia intermedia without serious clinical symptoms. As the main problem in β-thalassemia is the excess amounts of α-globin chains, people who have simultaneously developed some form of α-thalassemia along with β-thalassemia end up having less α-globin chains in excess. This is also the case in people with increased HbF production in whom γ-globin chains bind some of the excessive α ones. In cases like these, people have less severe anemia, although they also have less hemoglobin than "normal." This suggests that the model in Figure 4.4a does not work for all cases. Heterozygotes could have almost "normal" or totally "abnormal" levels of hemoglobin, as shown in Figure 4.4b.

One GWAS involving 235 mildly and 383 severely affected patients identified 23 SNPs in three independent genomic regions as being significantly associated with the severity of the disease. Not surprisingly, the highest association was found with SNPs within the *HBB* gene. The second most significant association was found with two genes on chromosome 6, the *HBS1 L* (HBS1 like translational GTPase) gene that encodes an enzyme, and the *MYB* (MYB proto-oncogene) gene that encodes a protein that regulates transcription. Both of these proteins are involved in the processes of production of blood cells. A third region was within the *BCL11A* (B-cell CLL/lymphoma 11A) gene on chromosome 2 that encodes a protein that seems to be implicated in leukemia. Various molecular changes, including single nucleotide changes as well as more extensive changes such as deletions, have resulted in more than 300 alleles of the *HBB* gene that are related to β-thalassemia. Most β-thalassemias are due to point mutations in the *HBB* gene. Several kinds of deletion are also associated with β-thalassemia. Some of them are restricted within the *HBB* gene, whereas others are more extensive, removing part of this gene and other sequences, thus resulting in no production of β-globin chains. Other deletions remove regulatory sequences of the *HBB* gene, leaving the gene intact but resulting in its decreased expression. There exist more than 200 point mutations related to

(a)   Parents

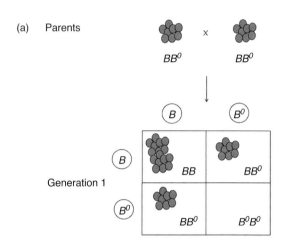

(b)   **Genotype**   **Phenotype**

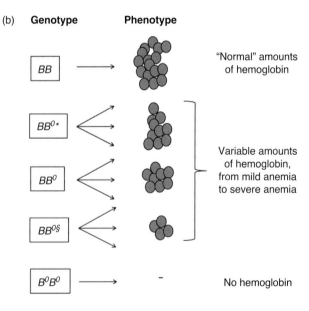

Figure 4.4   (a) The inheritance of β-thalassemia is usually described as that of a typical Mendelian characteristic. According to this, two parents who are carriers of the

β-thalassemia in regulatory sequences, in sequences related to splicing, within the *HBB* gene sequence, as well as various deletions.

All the above might make one think that *HBB* is the "gene for" β-thalassemia. However, the condition does not only depend on this gene. Carrying the disease-related alleles of *HBB* does not entail that one will definitely have the disease, as it is important to also consider other genes. It seems that people who have simultaneously developed a form of α-thalassemia (which is due to mutations in genes *HBA1* and *HBA2* on chromosome 16 that limit the production of α-globin chains), or who have an increased production of hemoglobin F (normally produced in fetuses) as γ-globin chains bind some of the excessive α-globin chains, eventually have less α-globin chains in excess and therefore less severe anemia.

Let us for simplicity call the alleles that produce "normal" amounts of these proteins α and β, and the mutated ones $\alpha°$ and $\beta°$. It is possible that two parents each carry one $\beta°$ allele, thus having reduced amounts of β-chains that nevertheless do not significantly impact their lives. It is also possible that one of their children will come to have two $\beta°$ alleles and thus develop a severe form of β-thalassemia. However, it is possible that the mother also carries an $\alpha°$ allele. This woman has fewer α- and β-chains than normal, and a reduced amount of hemoglobin, but she faces no significant problems because there is no excess of globin chains. Now, whereas each one of the offspring of this couple has a 25 percent probability of inheriting two $\beta°$ alleles (Figure 4.5), the probability for each one of them to inherit the $\alpha°$ allele is 50 percent (similar to Figure 4.4). These two events are independent, so in order to estimate the probability that one of the offspring would have two $\beta°$ alleles

---

Caption for Figure 4.4 (cont.)

defective allele have 25 percent probability of having an offspring who is homozygous and has the disease (circles represent the proportion of hemoglobin molecules; $\beta°$ is a defective allele and β is a "normal" one). (b) However, this model is insufficient to describe the phenotypic heterogeneity of the disease, as different heterozygotes can have a variety of phenotypes, from almost "normal" to almost defective ($\beta\beta°$, $\beta\beta°*$, $\beta\beta°§$ denote heterozygotes with different alleles).

Parents

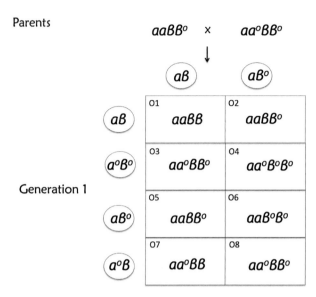

Figure 4.5   The father carries the β° allele and can produce two types of spermatozoa: those that have the disease-related allele and those that have the "normal" one. The mother also carries the β° allele as well as the α° allele. Therefore, theoretically, she can produce four kinds of ova: having both of these alleles, one of them, or none of them. As a result, several different kinds of offspring are possible: some will be completely healthy (e.g., O1); others will be carriers of either the α° or the β° allele (e.g., O7, O2); and some will likely have β-thalassemia (e.g., O6); but there will also be some who may not have significant problems because they have both the α° and the β° allele (e.g., O8).

and an α° allele, we have to multiply the above probabilities: $0.25 \times 0.50 = 0.125$, or 12.5 percent. Therefore, finding that a child carries two β° alleles does not entail that they will develop the disease. This could be the case for child O6, but child O4 might have mild anemia because of the simultaneous presence of the α° allele (Figure 4.5). Because some homozygotes for the β° allele might have β-thalassemia major, and others with exactly the same genotype might have a milder form of β-thalassemia intermedia, thinking of *HBB* as the "gene for" thalassemia is insufficient to give the whole picture.

Let us consider another disease. Familial hypercholesterolemia is a condition characterized by high amounts of cholesterol in blood, which results in coronary heart disease. By the 1950s, it had been shown that fatal events increased in proportion to increased blood cholesterol levels. For example, comparative studies of Japanese men living in Japan, Hawaii, and California showed that fat intake, plasma cholesterol, and the number of heart attacks all increased along the way from Japan to the USA. Around that time, it also became clear that there is an association between blood lipoproteins and the risk for heart attacks. It was soon shown that the plasma of people who had undergone a heart attack was characterized from increased levels of cholesterol-carrying lipoproteins – the low-density lipoproteins (LDL) – compared to individuals matched by age and sex without coronary heart disease. LDLs consist of several hundreds of hydrophobic (do not dissolve in water) molecules (cholesteryl ester) surrounded by a hydrophilic (dissolve in water) coat and a large protein called apolipoprotein B (apoB). LDL particles are considered to penetrate the walls of coronary arteries, where they become oxidized, get inside white blood cells, and convert them to cholesterol-laden foam cells. These initiate an inflammatory reaction, which makes the muscle cells of the arteries proliferate. In this way, what is described as atheromatic plaque grows (the phenomenon is described as atherosclerosis). At the same time, lipids derived from dead and dying cells accumulate in the central region of a plaque. When the plaque ruptures, a blood clot is formed that blocks the artery, thereby causing a heart attack.

Familial hypercholesterolemia and increased LDL levels result from mutations in the gene for the low-density lipoprotein receptor (LDLR) on chromosome 19. This is a transmembrane protein that removes LDLs from the blood by binding to their apoB protein. Mutations in the *LDLR* gene result in the reduced uptake of LDLs, which in turn leads to the accumulation of LDLs in plasma, tendons, and skin, and thus contributes to atherosclerosis. At the same time, because of limited or no LDL uptake, cells increase their own production of cholesterol by several times. Patients can be homozygous or heterozygous for this condition, and coronary heart disease occurs after a threshold of LDL exposure is reached – in early childhood in homozygotes and in early middle age in heterozygotes. The severity of the condition depends on whether the function of the LDL receptor is entirely disrupted or just reduced. The standard therapy for this condition is the administration of statins, which block cholesterol synthesis and thus lower cholesterol levels inside the cell, activate the synthesis of LDL receptors, and eventually lower the levels of LDL in plasma. Heterozygotes generally respond better to this kind of treatment than homozygotes, but statin treatment can be quite effective even in homozygotes if it starts before atherosclerosis develops. However, homozygotes are rather rare, with a frequency of less than one in a million people. More than 1,300 independent *LDLR* variants have been reported, more than three-quarters of which are due to base substitutions or small rearrangements in the exons. The rest of the changes include larger rearrangements, variants in the introns, and some promoter variants.

However, the *LDLR* gene is not the only one related to familial hypercholesterolemia. There is another gene, the *PCSK9* (proprotein convertase subtilisin/kexin type 9) gene on chromosome 1, which encodes a protein that is highly expressed in the liver and mediates the degradation of the LDL receptor molecules. The PCSK9 proteins bind to LDLRs in liver, promote their degradation and thus disrupt the process of recycling that returns them to the surface of the cell. Therefore, the number of LDLRs in the liver decreases and thus the amount of LDL in the blood rises. However, it has been found that some people with certain variants of the *PCSK9* gene seem to be protected against heart attacks. In a study, among 3,363 African Americans examined, 85 had non-sense mutations (ones that result in a stop codon and thus in a protein shorter than anticipated) in the *PCSK9* gene, which were associated with a 28 percent reduction in mean LDL cholesterol and an 88 percent reduction in the risk of coronary heart disease. Among 9,524 European Americans examined, 301 had a sequence variation in *PCSK9* that was associated with a 15 percent reduction in LDL cholesterol and a 47 percent reduction in the risk of coronary heart disease. Therefore, particular mutations in *PCSK9* can be protective. This is not the whole story, of course, as GWAS have identified associations between several other genes and coronary heart disease. This suggests, again, that thinking in terms of a "gene for" hypercholesterolemia or coronary heart disease is insufficient to give the whole picture. As shown in Figure 4.6, two siblings with the same *LDLR* alleles may have a very different phenotype because they carry different *PCSK9* alleles.

All the above point to an important distinction: It is one thing to have a mutation related to a disease and another to have the disease itself. Whereas patients suffering from β-thalassemia have one or more mutations related to the disease, there are many people who do not have the disease, although they have a relevant mutation. Therefore, even in these classic and exemplar cases of monogenic diseases there is heterogeneity that makes the notion of a "gene for" β-thalassemia or familial hypercholesterolemia difficult to sustain. This does not question the connection between the gene and the disease, as patients will most likely have one of the documented mutations. But one can question the inference made from a genetic test finding such a mutation that the child-to-be will definitely suffer from the disease. The disease supports the inference to the gene and can be explained by it, but the gene does not support the inference to the disease and does not necessarily predict it.

Multifactorial Diseases: Cancers

The term "cancer" is used to describe a variety of diseases, which neverthe-less have some common features. Cancers (also called neoplasms) result from the alteration of normal development and of tissue-repairing processes during one's life. They can be solid tumors that develop in organs such as breasts, colon, and lungs, or fluid tumors like those developing in blood during leukemia. A tumor is practically an overgrown mass of one's own cells. The problem tumors cause is that they consume so much of the resources of the body that the cells of the organ or tissue in which they develop die out, with bad consequences for the whole organism. When tumors are restricted to a tissue and grow slowly, they are not life-threatening and are called "benign." In contrast, when they are invasive, moving from one tissue to another, they are life-threatening and are called "malignant." In some cases, removal of the tissue is possible (e.g., of skin, colon, breasts), and if the whole tumor is removed then the individual can live without problems. However, there are certain organs where such a removal is difficult or impossible (e.g., lungs, stomach, bone marrow), with the unfortunate outcome of death.

Not surprisingly, given what I have presented so far in this book, a reductionist perspective of cancer has been dominant – but not the only one. In this view, which has been described as the somatic mutation theory

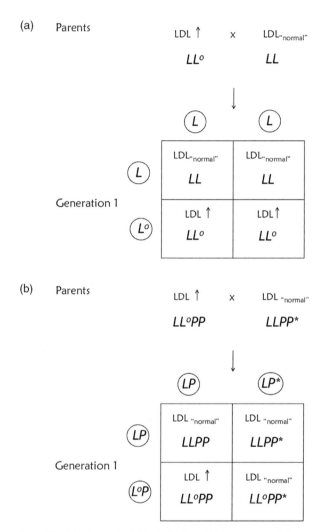

Figure 4.6 (a) When an individual with hyperlipidemia, who is heterozygous for a defective *LDLR* allele (*LL°*), and a "normal" individual have offspring, the

(SMT) of cancer, the vast majority of cancers are due to mutations, which in turn are due to exposure to carcinogens either before or after birth. Carcinogens can be physical (e.g., α, γ, X-rays), chemical (e.g., tobacco, smoke, DDT, asbestos), or biological (e.g., viruses) factors. In principle, the development of cancers such as colorectal cancer, melanoma, and esophageal cancer are associated with exposure to environmental factors. These cancers are described as sporadic. In contrast, inherited cancers are due to mutations inherited from one's parents, and they are the minority of all clinical cancers. Such cancers include retinoblastoma, breast cancer, ovarian cancer, and familial colon carcinomas. However, there are no "genes for" cancer, although it is certainly the case that particular genes are associated with particular cancers. These are the proto-oncogenes that generally promote cell division and the tumor suppressor genes that generally inhibit cell division. Mutations in proto-oncogenes change them to oncogenes and result in a gain of function, thus enhancing cell division; mutations in tumor suppressor genes result in a loss of function, thus also enhancing cell division. Thus, tumor development in humans comprises multiple steps, each of which is characterized by alterations that progressively transform healthy human cells into malignant ones.

It is important to note that when mutations are found in tumors, it is not clear which ones had a causal influence on the development of cancer (driver mutations) and which ones were irrelevant because the cancer had already developed (passenger mutations). It has been found that most somatic mutations in cancer cells are likely to be passenger mutations; however, about one in five genes are estimated to carry a driver mutation and therefore may

---

Caption for Figure 4.6 (cont.)

probabilities for each of their children to have hyperlipidemia is 50 percent. All of their offspring with an *LL°* genotype would be expected to have high LDL levels in blood. (b) However, it is possible for an *LL°* individual not to have hyperlipidemia. This cannot be explained by the *LDLR* genotype, but only if one considers the effect of the *PCSK9* alleles. Individuals with *PP* genotypes can have problems. However, if a person carries an allele with a loss-of-function mutation in the *PCSK9* gene (*P**), this allele can moderate the effect of the *LL°* genotype. Thus, this person does not have hyperlipidemia, as the LDL levels are within the normal range.

function as genes promoting cancer. Driver mutations are causally implicated in the development of a tumor (oncogenesis) by conferring a growth advantage on cancer cells that then proliferate faster than their neighboring cells. In contrast, passenger mutations do not confer any growth advantage, and thus do not contribute to the development of cancer. They exist only because they just happened to occur. Distinguishing between driver and passenger mutations is a major challenge. However, driver mutations are usually found in particular cancer-related genes, whereas passenger mutations are more or less randomly distributed in the genome.

In this view, cancer development is a process analogous to natural selection; it is based on the continuous acquisition of mutations in individual cells and the subsequent selection process that takes place among the phenotypically different cells. The first process results in what are described as somatic mutations (i.e., mutations occurring anew in the body). These mutations include all kinds of DNA sequence changes, as described in Figure 3.8. The rate at which such mutations occur increases in the presence of carcinogens. Cells may also acquire DNA sequences from exogenous sources such as viruses. Other processes, such as problems in the repair mechanisms of DNA, may further increase the number of newly occurring mutations. Among the genetically diverse cells thus produced, a selection process may take place. Cells with deleterious mutations may die out, but cells carrying mutations that confer the ability to proliferate and survive more efficiently than their neighboring cells may also increase in numbers. Some of these cells may just remain invisible or develop into benign structures such as skin moles, but occasionally some may become cancer cells that are able to proliferate autonomously and invade other tissues.

Cancer researcher Bert Vogelstein and his colleagues were the first to demonstrate that colorectal tumors result from the gradual accumulation of mutations in specific oncogenes and tumor suppressor genes. Vogelstein and colleagues have recently suggested that most human cancers are caused by 2–8 sequential mutations that take place over the course of 20–30 years. Each of these mutations provides, directly or indirectly, a selective growth advantage to the cell in which it exists. Evidence suggests that there are approximately 140 genes in which mutations that contribute to cancer occur. These genes affect several signaling pathways that regulate three important cellular processes: cell fate determination, cell survival, and genome maintenance. The pathways affected in different tumors are similar, but the mutations in each individual tumor are different, even between tumors of the same type. This is described as genetic heterogeneity of cells of the same tumor, and as a result of this there may be different responses to cancer therapy. Vogelstein and colleagues have also suggested that the lifetime risk of several types of cancer is strongly correlated with the total number of cell divisions that normally renew a given tissue. This means that many cancers develop in particular tissues because of random mutations occurring during DNA replication in normal cells and not because of external, carcinogenic factors. This explains why cancers develop in some human tissues a lot more often than in others: Two-thirds of this variability is explained by newly occurring mutations, as the more the cells in a tissue divide, the more likely it is for mutations to occur and thus for cancer to develop.

Cancer biologists Douglas Hanahan and Robert A. Weinberg have argued that most cancers exhibit six important physiological changes, which are acquired during the development of the tumor and which allow cells to overcome certain anticancer defense mechanisms that they naturally exhibit. They have described these as the hallmarks of cancer: (1) self-sufficiency in growth signals; (2) insensitivity to antigrowth signals; (3) evasion of programmed cell death (apoptosis); (4) limitless replicative potential; (5) sustained angiogenesis; and (6) tissue invasion and metastasis. Collectively, all these contribute to the uncontrolled proliferation of cancer cells and the development of the tumor. More recently, Hanahan and Weinberg have argued that research has revealed two more emerging hallmarks that have the potential to be as important as the other six: reprogramming of energy metabolism and evading immune destruction. Therefore, despite the differences in the more than 100 types of cancers that have been described, they are considered to share a common pathogenesis. In this view, normal human cells become cancerous after many of the aforementioned hallmarks occur successively, and only late-stage cancers exhibit all of them. This is the outcome of successive mutations that happen in the same or neighboring cells, which eventually become cancerous ones. Each time a certain hallmark appears, it becomes more likely that another one will occur next because cells become more disorganized. Thus, by continuously having more of such disorganized cells, it becomes more likely that further disorganization will take place.

This brings us to a very important distinction, that between a genetic disease and an inherited disease. Whereas an inherited disease is quite often genetic as well, there exist numerous genetic diseases that are not inherited. Cancers are an exemplar case in which there may be a genetic change (i.e., a mutation in DNA) that was not inherited from one's parents (it was not a germline mutation), but that happened anew (de novo) in one's somatic cells (and this is why it is called a somatic mutation). Even in the cases of cancers that run in families, it is not always the case that they are due to genetic inheritance. In contrast, it is possible for family members to have mutations leading to cancer because they were exposed to the same carcinogen. We tend to think (and also to be taught at school occasionally) that we inherit our DNA from our parents in some fixed form. However, it has been shown that offspring often carry new mutations that do not exist in their parents and that are therefore novel. Using different methods, studies have estimated the mutation rate in humans to be 1.2×10^{-8} per nucleotide per generation, and $1.4–2.3 \times 10^{-8}$ mutations per nucleotide per generation. Details notwithstanding, these data suggest that any individual is likely to carry at least 72 new mutations ($1.2 \times 10^{-8} \times 6 \times 10^9 = 72$). This is why a genetic disease may not be due to some mutation inherited from one's parents, but due to a new one.

The view that emphasizes the impact of mutations on cancer has been criticized by cancer researchers Carlos Sonnenschein and Ana Soto. They have questioned and criticized the validity of the hallmarks of cancer proposed by Hanahan and Weinberg and they have argued that the SMT is insufficient for explaining cancer. After criticizing all hallmarks one by one, Sonnenschein and Soto concluded that the SMT has not contributed to an increased understanding of cancer and has not driven a meaningful beneficial translation of the respective research findings for curing cancer. In contrast, they have proposed what they have called the tissue organization field theory (TOFT). This theory differs from SMT in at least three important aspects:

1.   Cancer is a tissue-based disease; it is not a cell-based one. According to the SMT, cells are the targets of carcinogens, and therefore research

should focus on events inside cells. In contrast, the TOFT focuses on the interactions among the various different cell types.

2. Proliferation is the default states of all cells; it is not the case that cells require some kind of (internal or external) stimulation in order to proliferate. The SMT assumes that quiescence is the default state of cells in multicellular organisms, and that therefore cells require stimulation by either "oncogenes" or "growth factors," in order to proliferate. In contrast, according to the TOFT, proliferation is the default state of all cells, and the tissues in which they reside constrain their ability to proliferate and move. When carcinogens loosen the tissue constraints, cells can regain their default state and proliferate again, eventually forming tumors.

3. What is disrupted by the carcinogenic factor are development processes; it is not the DNA that is disrupted. According to the TOFT, cancer is due to abnormal tissue interactions that make developmental processes go awry, and not due to mutations, as the SMT suggests. Mutations are therefore just an epiphenomenon of cancer, not its main causes.

In short, whereas the SMT perceives cancer as a genetic disease, the TOFT perceives cancer as a developmental disease. Conceptually, and in line with what I have written in this book, I would tend to favor the TOFT. In many cases of cancers, there exist alterations in the structure of DNA or chromosomes, but it is not always clear whether these are a cause or an effect of cancer. Philosopher of science Anya Plutynski has suggested that rather than two competing theories, what we see here can be "better characterized as a gradual shift in understanding and assimilation of novel ideas." These novel ideas seem to point in the same direction: cancer is not only about mutations.

An important conclusion from this chapter is that the common distinction between "simple" and "complex" or "monogenic" and "multifactorial" characteristics or disease is an oversimplification. Whereas it is certainly the case that some characteristics and some diseases are strongly associated with one gene, other DNA sequences may also be involved in their development. In addition, it is possible that numerous variants exist within the same gene and have a variety of phenotypic outcomes. Therefore, even in the case of "simple" or "monogenic" characteristics and disease, the path from the DNA sequence to the associated phenotype can be quite complex. As the examples in this chapter have shown, there are several reasons for this, such as that variation within a single gene can result in very different phenotypes, that phenotypes cannot be accurately predicted from genotypes, and that several genetic and nongenetic factors may also have a contribution to the respective phenotypes. Therefore, if many genes produce or affect the production of a protein that in turn affects a characteristic or a disease, it makes no sense to identify one gene as the "gene for" this characteristic or disease. Which begs the question: What is it, then, that genes do?

5 What Genes "Do"

The Development of Characteristics in Individuals

To understand what genes "do," we have to consider what happens during development. The first and most striking evidence that the local environment matters for the outcome of development was provided by the experiments of embryologists Wilhelm Roux and Hans Driesch in the late nineteenth and early twentieth centuries. Roux had hypothesized that during the cell divisions of the embryo, hereditary particles were unevenly distributed in its cells, thus driving their differentiation. This view entailed that even the first blastomeres (the cells emerging from the first few divisions of the zygote – that is, the fertilized ovum) would each have different hereditary material and that the embryo would thus become a kind of mosaic. Roux decided to test this hypothesis. He assumed that if it were true, destroying a blastomere in the two-cell or the four-cell stage would produce a partially deformed embryo. If it were not true, then the destruction of a blastomere would have no effect. With a hot sterilized needle, Roux punctured one of the blastomeres in a two-cell frog embryo that was thus killed. The other blastomere was left to develop. The outcome was a half-developed embryo; the part occupied by the punctured blastomere was highly disorganized and undifferentiated, whereas those cells resulting from the other blastomere were well-developed and partially differentiated. This result stood as confirmation for Roux's hypothesis.

However, the experiments that Driesch conducted with sea urchin embryos a few years later pointed to a different conclusion. Whereas it was impossible to separate the blastomeres of a frog embryo without killing it, sea urchin blastomeres could be separated completely just by shaking. So, what Driesch did was to apply the same procedure with Roux, and then shake a container in which two-cell urchin embryos existed. This resulted in the separation of the blastomeres of each pair, which Driesch let develop on their own. Whereas he expected each of the blastomeres to develop to a half-embryo, as Roux's experiments had suggested, he soon observed that each blastomere had developed to a complete, albeit a bit smaller, embryo. This showed that each of the blastomeres of a two-cell embryo had the potential to develop to a complete embryo if it were separated from the other one. This could not be possible if hereditary information was distributed differently in each of the blastomeres. Driesch also observed that if the two blastomeres developed together, each of these would give rise to cells that would form a different part of the embryo. The important conclusion was that the local environment has an impact on the outcome of development. When a blastomere is on its own, it gives a complete embryo; when it is attached to another blastomere, no matter whether the latter is dead or alive, it gives a half-embryo. This happens because development depends on signals from the local environment that make cells move and differentiate in a coordinated way.

Evolutionary geneticist Richard Lewontin has suggested that a major confusion arises from the original sense of the term development itself. "Development" literally means unrolling something that already exists somewhere. It is no coincidence that the same term was used for the process of printing on paper the image that already existed in a photographic film. During the eighteenth century, there were two major competing theories of development: preformationism and epigenesis. Preformationism theories suggested that a germ capable of development already possessed a certain structure that somehow preconditioned the adult form. In contrast, the theories of epigenesis suggested that a germ capable of development was unformed. Lewontin has argued that it is actually the preformationism view that has dominated genetics. Of course, nobody has thought that an adult form is preformed in the first cells from which an organism develops. Nevertheless, the idea of a "blueprint" that contains the necessary information for the production of an adult form is quite similar, because it accepts that there exists some fixed genetic essence inside the organism that causally

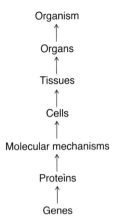

Figure 5.1 According to the "blueprint" metaphor, genes contain the necessary information for the production of an adult form, based on the assumption that there exists some fixed genetic essence inside the organism that causally determines its form, which in turn is best explained in a reductionist manner. Causality thus proceeds from genes onward.

determines its form, which in turn is best explained in a reductionist manner (as shown in Figure 5.1). The "blueprint" metaphor thus encompasses all major misunderstandings about genes: determinism, essentialism, and reductionism. In this sense, the role of the external environment is limited to certain conditions that may be necessary to trigger the developmental process and to allow it to proceed along a rather predetermined path. This is the interpretation of many scholars, who we met in Chapter 1.

Rhetorical criticism scholar Celeste Condit, who we already met in Chapter 1, has argued for a different interpretation of the blueprint metaphor "as a preliminary outline, to which other elements would be added and which might be modified by the material limitations constituted by those other elements." In her view, the blueprint metaphor can be interpreted as a probabilistic and malleable outcome, rather than as a deterministic one, as an initial material that can be molded and transformed in various ways, rather than as something whose final form is already predetermined. Condit actually tried to find empirical evidence for this interpretation by conducting a short study with 137 US college students who read sample genetics news articles and were asked for their interpretations of the "blueprint" metaphor. Among them, 58 provided responses that were explicitly nondeterministic and 39 provided explicitly deterministic ones, whereas the others provided mixed or irrelevant responses. The nondeterministic responses relied on interpretations of the "blueprint" metaphor that referred to genes as operating in a partial and probabilistic fashion, as well as being malleable and not determining one's destiny.

However, a more recent study that explicitly contrasted the "blueprint" metaphor with the "instruction" metaphor gave different results. In this case, 324 US adults were given a definition of genes that used either an "instruction" metaphor or a "blueprint" metaphor. The "instruction" metaphor suggested that

> A unit of DNA is called a "gene." Every person has the same 30,000 genes, but a person can have different versions of these 30,000 genes in comparison to others. Each gene provides instructions for a specific chemical substance that the body uses. Not all instructions are followed all the time. As with any set of instructions, how they are followed or whether they are followed depends on many factors.

In contrast, the "blueprint" metaphor suggested something less flexible:

> Genes are working subunits of DNA. DNA is a vast chemical information data base that carries the complete set of instructions for making all the proteins a cell will ever need. Each gene contains a particular set of instructions, usually coding for a particular protein.

The researchers found that the "blueprint" metaphor promoted stronger essentialist beliefs that genes are "a causal, powerful, deterministic driver of health," compared to the "instruction" metaphor.

I would also argue that the blueprint metaphor had best be avoided. Here is why: The development of tissues and organs is not controlled by genes or DNA only, but also by the exchange of signals among cells. These signals consist of gradients of signaling proteins: "a lot of" protein A might mean "do this"; "some" protein A might mean "do that"; and "no" protein A might mean "do nothing at all." Details notwithstanding, what is important to note is that whatever a cell does and whatever kinds of signals it sends out depends on the kind of signals it receives from its immediate environment. Therefore, neighboring cells are interdependent, and it is local interactions among cells that drive the developmental processes. At the same time, these localized processes also make the development of different organs relatively independent. This allows for control and changes in each organ independently from other organs. Most interestingly, this is often achieved by using and reusing the same signaling proteins. To use a simplistic metaphor, the route and the destination of a car are not determined by the place the driver wants to go; they are determined by the signals the driver receives and his/her reaction to them. Thus, the driver will stop at a red light, will turn to avoid another car, and may change route to avoid traffic. Eventually, the driver may arrive at the intended destination; but the driver may also decide to return home because of bad weather conditions. This metaphor shows that the destination and the route may change because of the signals the driver receives and do not only depend on where they want to go. In this sense, environmental signals matter for development.

Development comprises two major phenomena: growth and differentiation. Growth consists in cell proliferation, and thus the total number of cells of the body increases and the body grows. Which cells will and will not divide is determined by local signals and interactions. At the same time, cells also differentiate and give rise to the various tissues and organs of the body. Perhaps the most astonishing stage of this process is gastrulation, during which there is a massive reorganization of the embryo from a simple spherical ball of cells, the blastocyst, to a fetus consisting of different types of layers: endoderm, mesoderm, and ectoderm. Epithelial tissues and glands of the digestive and respiratory systems are derived from the endoderm; connective tissues (cartilage, bone, blood) and muscle (cardiac, skeletal, smooth) are derived from the mesoderm; and skin and parts of the nervous system are derived from the ectoderm. The process of differentiation is also determined by local signals and interactions. The combination of the processes of growth and differentiation make an early embryo with radial symmetry to develop to a much larger and more complex fetus with bilateral symmetry (Figure 5.2).

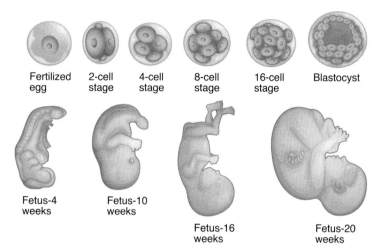

Figure 5.2 Human embryo development. A spectacular change in form takes place in just a few weeks, as an embryo with radial symmetry gradually develops to a fetus with bilateral symmetry (© BlueRingMedia).

During development, cells multiply, differentiate, and migrate to various parts of the developing embryo in a coordinated manner. To understand this, imagine a very large number of humans moving within a limited area in such a way that no one steps on anyone else. How is this possible? There are two plausible explanations. One is that someone organized the crowd and told them how to move in a coordinated way, as would be the case in a military parade. Another is that people simply respond to local cues (see, listen) and avoid stepping on one another, as would be the case while walking in a commercial street on the first day of sales (e.g., walking along Oxford Street on Boxing Day). We now know that the organization of cells in the developing body is achieved in the second way. There is no centralized coordination of development; cells respond to signals from their local environment. Of course, something might go wrong: people might for some reason panic and start stepping and falling onto one another. Under normal conditions, though, development takes place on the basis of local signals. But how do genes fit in this picture? Signal production, signal reception, and signal response all depend on proteins, which in turn are produced by the expression of genes. Therefore, genes are implicated in this unconscious coordination of development, but they in no way determine its course and its outcomes, which include characteristics and diseases.

An appropriate way to conceptualize the role of genes in development is to think of an organism as an origami (Figure 5.3), as suggested by developmental biologist Lewis Wolpert. According to this, the DNA of the fertilized ovum is not a blueprint that contains a full description of the adult organism that will emerge from development. Rather, it contains a set of instructions for making the organism, which will affect cell proliferation and differentiation. These instructions are about "how to make" the adult organism, not about "how the organism will look." Therefore, the DNA of the fertilized ovum contains a generative plan, not a descriptive plan. What matters according to this analogy is that what is available are the instructions about when, where, and how to fold the paper in order to make a structure. A description of how the origami structure will look would be entirely useless because it would provide no clues (at least not clear ones) about how to generate it. In the same sense, a description of how the adult form will look is useless; what is needed is a set of instructions about how to generate it. Therefore, what happens during human development is not that genes containing the blueprint for the adult human express themselves and thus the organism is constructed to resemble this blueprint. Rather, cells follow the generative plan encoded in genes and the signals they receive from their environment. It is from the combination of numerous local signals coming from the intracellular and the intercellular space that cell division, proliferation, and differentiation take place during development. Thus, from a single fertilized ovum an organism develops. Appropriate signals will drive the production of the anticipated "normal" outcome, whereas "bad" signals can make things go wrong and bring about developmental defects.

A note on metaphors: What is important in this analogy is the difference between "descriptive" and "generative," which I represented as the difference between "how something will look" and "how to make something." In both cases I referred to a "plan," even though in the original origami analogy Wolpert actually used the term "program." In my view, the term "program" implies both instructions and their implementation, whereas the term "plan"

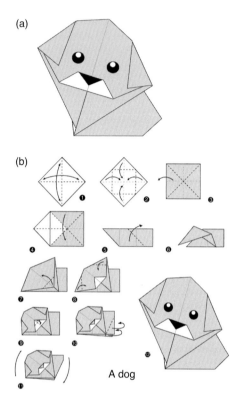

Figure 5.3 (a) A descriptive plan; (b) a generative plan. What is encoded in DNA should be conceived as a generative plan, not a descriptive one. (© tofang).

is about instructions only. The notion of a genetic program can be very misleading because it might be perceived to imply that, if it were technically feasible, it would be possible to compute an organism by reading the DNA sequence alone.

Here is another way to conceptualize the role of genes and DNA in development. Imagine a famous play such as, my favorite, *The Phantom of the Opera*. This was originally a novel by Gaston Leroux, published in 1909–1910.

However, it became famous as a musical that has been a huge success since 1986 in London and 1988 in New York, based on a play and music by Andrew Lloyd Webber. Let's imagine that he used the same script for the London and the New York theater plays, and that he also sent it to several secondary school, college, and university theater groups in the UK and the rest of the world. Although the script (DNA) is exactly the same, the play (phenotype) can be quite different in its various versions. The reason for this is that in each theater group (developmental system) there is a different director and different actors (proteins) who might or might not vary in how the script (DNA) is interpreted (expressed). One group might follow the script exactly as written; another might make minor changes in the words of Christine; another one might change the words and the time of appearance of the Phantom (the copyright license might not actually allow such changes, but let's assume that this is possible). Such changes might result in variations of the same play – the differences between the New York and the London play might be insignificant because of the professionals involved, but you can imagine how creative high school and college students can be, and therefore that it is possible for them to further increase the variation in the final outcome (how happy Andrew Lloyd Webber would be with that need not concern us here).

It is in this sense that what is written in *The Phantom of the Opera* script is not a descriptive plan (e.g., how the play should look) but a generative one (e.g., how many actors are needed, what each one of them has to say and when, etc.). Not surprisingly, the play that the official London, the official New York, a Cambridge secondary school, and an Oxford University theater group would produce based on the same script might differ significantly, as it would depend on who was the director, who read the script, how that person selected the actors and assigned the roles, what instructions that person gave to them, and what changes to the original script, if any, that person made. All plays would be *The Phantom of the Opera* in the broader sense, but with differences. What is important is that these differences would be due to differences in the implementation of the script and not in the script itself. In the same sense, the development of an organism should be conceived as the implementation of a plan included in the DNA of the fertilized ovum. Variations in the outcome of development could exist even if there were no variations in the sequence of DNA, because of the unique interactions between an organism's genome and the external environment(s) in which it lives. Thus, as different plays are possible based on the same script, it is also possible that individuals with exactly the same genotype may have different phenotypes because of environmental influences during development.

The reason for this, and this is the most important point, is that it is not genes but cell interactions through signaling molecules that guide development. How a cell responds to a particular signal depends on its internal state, which in turn reflects its developmental history. Therefore, different cells might respond to the same signal in different ways, and the same signal can be used in various ways in the developing embryo. One of the most striking examples showing how development depends on communication among cells and responses to local signals stemming from a generative plan, rather than on the local implementation of a larger descriptive plan, is the inter-action of genes and the environment in brain development. The genes that are active in the brain cells during its development do not specify its final structure (e.g., which connections will be made between neural cells and how many of them in each case). Rather, the proteins specified in those genes together make mechanisms that strengthen, weaken, or destroy connections according to the received signals. These mechanisms are constantly refined, thereby changing the way that signals associate with one another. Therefore, the particular neural connections are the outcome of the interaction between certain proteins – encoded by genes – that participate in mechanisms that make neural connections, and their environment. These proteins do not produce predetermined outcomes; if they actually did so, then nothing could be accommodated through learning and experience. Thus, either "bad" genes or "bad" environment can result in mental deficiencies, whereas "good" genes can contribute to a healthy mind only in an environment that provides the stimuli that will lead to the appropriate internal connections during brain development.

Genes are Implicated in the Development of Characteristics in Individuals

It should by now be clear that what matters is not only whether a gene is expressed, but also in which context the gene product does whatever it is that it does. This context, in turn, can be influenced by the immediate cellular and the broader tissue–organismal environment. Genetic determinism assumes that a gene determines a phenotype, and that the environment in which the cell or the organism lives makes no difference. This can be the case in laboratory populations where the environmental conditions are controlled and are relatively uniform. However, this is not at all what happens in natural populations. This can become evident if we compare the different phenotypes of a particular genotype in different environments. Such a representation is called a reaction norm.

A classic example of how the same genotypes can actually produce different phenotypes in different environments is the experiment with *Achillea millefolium* plants. Seeds from 81 natural populations of *Achillea* were collected, germinated, and grown at Stanford. Then 30 individuals from 14 populations were cloned and planted at three different stations at Stanford (elevation 30 m), Mather (elevation 1,400 m), and Timberline (elevation 3,050 m). For each of these individuals, morphological and physiological characteristics were recorded over a three-year period. The results of this experiment showed variability in morphological and physiological characteristics among the three populations at the different elevations: there were different stem heights at different elevations. These results show that genetically identical individuals (assumed to have identical genotypes for all those DNA sequences related to stem height) do not necessarily have the same phenotype for the same characteristic in all environments. Therefore, the phenotype is not invariantly determined by genes; rather, it is the outcome of the interaction between genetic and environmental factors.

How is it possible for a given genotype to give such different developmental outcomes in different environments? This is absolutely natural. The problem is that this is not often discussed in genetics education and outreach. Development actually has two distinct and complementary aspects: robustness and plasticity. Developmental robustness is the capacity of individuals of the same species to exhibit the general characteristics of their species irrespective of the environment they live in, thus resulting in consistency of phenotype in different environments. For instance, plants of the same species will have similar stems and leaves. Developmental plasticity is the capacity of individuals of the same species with the same genotype to exhibit phenotypic variation, and thus to produce different phenotypes during development as a response to local environmental conditions. For instance, plants of the same species will not have stems and leaves of exactly the same size.

Let's consider again *The Phantom of the Opera*. I explained that different theater groups and directors following exactly the same script might end up with quite different plays. All plays would be *The Phantom of the Opera* (this is robustness); however, one of them might emphasize more the character of Christine, whereas another might emphasize more the character of the Phantom, thus diverging from Andrew Lloyd Webber's script (this is plasticity). Following the same script does not guarantee producing the same play if different directors and actors are involved. This is why not only the script (DNA), but also how directors and actors use it (development) matters. Of course, organisms are a lot more complicated than theater plays. They are dynamic systems characterized by molecular interactions of various kinds.

One might raise an objection at this point: It is indeed possible that some characteristics are plastic and therefore can be influenced by the environment. The height of the *Achillea millefolium* plants could indeed be different at different elevations. But could the type of flowers be affected in the same way? To take an example closer to home, think of our height and the color of our eyes. Everyone would probably agree that environmental factors, such as nutrition, could make a difference in height. Thus, it would be possible, even though it's less likely, that two identical twins had a different height because their nutrition differed significantly. But could the environment affect the color of their eyes? The answer is yes, even though eye color could be less likely to be affected by the environment than height. This does not mean that one could be born with brown eyes and end up having blue eyes in adulthood. Rather, it means that one may have the DNA sequences associated with brown color, but be born with a lighter color because for some developmental reason melanocyte formation in the eyes did not go as would be expected on the basis of DNA sequences alone (see Chapter 4). The important point made here is that genes indicate a potential or a predisposition; however, whether or not this is realized depends on the developmental processes that in turn depend on the environment. This is why the idea of genetic determinism is wrong.

Genes should not be considered as our essences, either. First, genes are not fixed entities that are transferred unchanged across generations. Changes that affect both the information that they encode and their expression can occur from one generation to the next. In addition, genes do not simply specify characteristics from which their existence can be inferred. For example, when *Arabidopsis thaliana* plants with the same genotype developed under different conditions, some being exposed to mechanical stimulation and others not, they came to have different characteristics. In particular, those plants stimulated by touch developed shorter petioles and bolts (a developmental response known as thigmomorphogenesis that may be an adaptive response to wind, precipitation, and/or attacks by insects). In this case, just seeing the different characteristics of these plants might make one infer that they had different genotypes, yet this conclusion would be wrong. Therefore, genetic essentialism is also wrong, as it is not certain that the existence of particular genes will result in particular phenotypes, and thus their existence cannot necessarily be accurately inferred by the observed phenotypes.

This phenomenon in *Arabidopsis thaliana* is a good illustration of robustness and plasticity. All plants exhibited the same general characteristics (robustness) but differed in the length of petioles and bolts as a response to environmental conditions (plasticity). What genetics research consistently shows is that biological phenomena should be approached holistically, at various levels. For example, as genes are expressed and produce proteins, and some of these proteins regulate or affect gene expression, there is absolutely no reason to privilege genes over proteins. This is why it is important to consider developmental processes to understand how characteristics and disease arise. Genes cannot be considered alone, but only in the broader context in which they exist (cellular, organismal, environmental). And both characteristics and disease in fact develop – they are not just produced. Therefore, reductionism, the idea that genes provide the ultimate explanation for characteristics and disease, is also wrong. In order to understand such phenomena, we need to consider influences at various levels of organization, both bottom-up and top-down (Figure 5.4).

All this shows that developmental processes and interactions play a major role in shaping characteristics. Organisms can respond to changing environments through changes in their development and eventually their phenotypes. Most interestingly, plastic responses of this kind can become

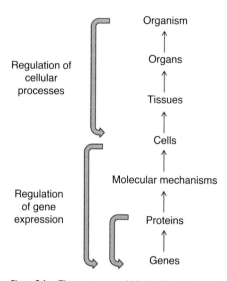

Figure 5.4   There are several kinds of interactions between the various levels of organization of organisms. Therefore, we need to consider not only what processes genes affect, but also what affects their own function.

stable and inherited by their offspring. Therefore, genes do not predetermine phenotypes; genes are implicated in the development of phenotypes only through their products, which depends on what else is going on within and outside cells. It is therefore necessary to replace the common representation of gene function presented in Figure 5.5a with others that consider development, such as the one in Figure 5.5b. Genes do not determine characteristics, but they are implicated in their development. Genes are resources that provide cells with a generative plan about the development of the organism, and have a major role in this process through their products. This plan is the resource for the production of robust developmental outcomes that are at the same time plastic enough to accommodate changes stemming from environmental signals.

Now, although I argue that genes do not determine characteristics, and that genes are neither our essences nor the ultimate explanations for the

(a)

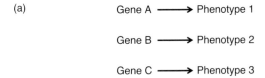

(b)

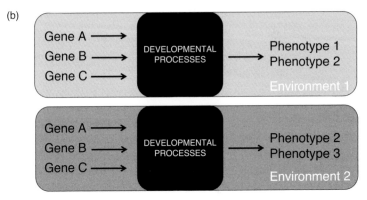

Figure 5.5   (a) A misrepresentation of gene function: a single gene determines a single phenotype. It should be clear by what has been presented in the present book so far that this is not accurate. (b) A more accurate representation of gene function that takes development and environment into account. In this case, a phenotype is produced in a particular environment by developmental processes in which genes are implicated. In a different environment, the same genes might contribute to the development of a different phenotype. Note the "black box" of development.

development of characteristics, it is indeed the case that changes in single genes can have a big phenotypic impact. But, whereas a change in a gene may even disrupt a whole developmental process, the final outcome is in no way determined by this gene alone. Let us consider the *SRY* (sex-determining region Y) gene on the Y chromosome as an example. As mentioned in Chapter 2, all humans have 23 pairs of chromosomes, and men and women differ in that women have two X chromosomes and men have an X and a Y chromosome. The default developmental outcome for the human embryo is to become a female. What makes the difference in what will happen is the expression of the *SRY* gene. This gene affects a pathway that guides the development of the male or the female sex (Figure 5.6). In this sense, *SRY* makes a difference for the development of sex. Embryos carrying the Y chromosome and the *SRY* gene develop testes and a male reproductive system, whereas those not having either the Y chromosome or the *SRY* gene develop ovaries and a female reproductive system. Interestingly, it has been found that a mutation (four-nucleotide deletion) in the *SRY* gene was adequate to make an XY individual develop as a female with underdeveloped reproductive organs, as well as that a translocation of part of the Y chromosome including the *SRY* gene onto the X chromosome was adequate to make an XX individual develop as hermaphrodite (carrying both male and female organs).

However, this does not in any way entail that *SRY* is the gene "for" maleness. As shown in Figure 5.6, several proteins (and therefore genes) are involved in the process of sex differentiation. The bipotential precursor of gonads (testes

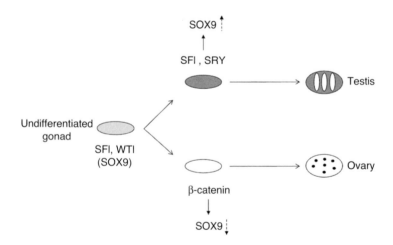

Figure 5.6   Sex differentiation in mammals. The SRY protein is only one of the proteins involved in this process.

and ovaries) is established by proteins, including SF1 and WT1, the early expression of which might also initiate that of SOX9 in both sexes; β-catenin can begin to accumulate at this stage, and in XX cells its levels could repress SOX9 production. However, in XY cells, increasing levels of SF1 activate the production of SRY that, with SF1, enhances SOX9 expression. If SRY activity is weak, low, or late, there is no SOX9 expression as β-catenin levels accumulate and shut it down. In the testis, SOX9 promotes the testis pathway, and it can do so even in the absence of SRY. Therefore, the *SRY* gene does not "determine" sex. However, it is indeed the case that mutations in this gene can have an enormous impact on the development of an individual. It is therefore changes in genes that can have an impact on developmental processes, and genes can act as difference-makers that affect the final outcome of development (this is discussed in detail in the next section).

Some syndromes also provide a striking illustration that the phenotype is not an outcome determined by genes but rather by complex interactions among gene products and cells. At some point during development, some cells in a structure that is called the neural crest are further differentiated to form the various structures of the face. But these cells require particular proteins in order to produce new cellular material and multiply, one of which is Treacle. Mutations in the gene *TCOF1* (treacle ribosome biogenesis factor 1) on chromosome 5 result in cell death. Therefore, very few cells exist to form the structures of the face, and the outcome is a condition in which individuals have slanting eyes, underdeveloped cheeks, a small lower jaw, drooping eyelids, and small or absent earlobes – the condition is known as Treacher–Collins syndrome. This syndrome is an example of the complexities that characterize development, and of the connections between genetics, development, and disease. Whereas in popular parlance stating that there is a gene for Treacher–Collins syndrome would make people think that there is a gene causing the respective facial abnormalities, this is not the case. The *TCOF1* gene is not one that is related to face construction but rather to cell proliferation. When Treacle is not operating properly, the neural crest cells do not have an adequate number of ribosomes and eventually become so stressed that they die. Consequently, a smaller number of cells are available for making the face.

The take-home message should be obvious: Genetics, and genetic inheritance, makes more sense in the light of development. Or, most distortions and misunderstandings of how genetic inheritance takes place have occurred because of the failure to take developmental processes into account. Genes are implicated in the development of characteristics and disease, but do not determine any of them in any way. Developmental processes show that genetic determinism, genetic essentialism, and genetic reductionism are inaccurate. It is therefore important to think of genes not as the ultimate determinants of who and what we are, but as the resources on which our cells draw while organisms develop and live under particular environmental conditions.

The Variation of Characteristics in Populations

Francis Galton is famous for many contributions to science and mathematics, as well to social policy through eugenics. One of these was the introduction of the contrast between nature and nurture in the context of heredity. Nature referred to the influence of one's biological makeup, whereas nurture referred to the influence of one's upbringing. Galton thus conceptualized nature and nurture as being associated with prenatal and postnatal influences, respectively. But he did not think of them as two distinct kinds of influences, one responsible for the transmission of characteristics (nature) and the other for their development (nurture). Rather, he thought of them as two necessary sets of influences acting simultaneously, but of which nature was always stronger. In principle, it would be very difficult to isolate and identify the influence of either nature or nurture, as individuals inherit both from their parents. For instance, was the son of a renowned musician a talented musician as well because of biologically inherited factors or because of the musical environment in which he was brought up? Galton tended to accept the former. To establish this, he relied on the study of "identical" twins. He thought that as these had the same nature but different nurture, they might provide the clues for distinguishing between the influence of each. Galton thus sent questionnaires to many parents of twins, aiming at establishing how their life experiences affected the similarities and differences between twins. The study of twins became the standard method for understanding the contributions of nature and nurture.

It is important to note that "identical" twins are not really identical, genetically speaking, as they do not really have 100 percent the same DNA for two reasons. The first is that, as mentioned in Chapter 4, a certain number of new somatic mutations occur in each of us. Therefore, it is possible that after the clusters of cells that will give rise to the twins are separated, certain mutations occur in one of them and not in the other. The resulting difference will be minor, but it exists. The second reason is that epigenetic changes – that is, changes in DNA and chromosomes other than changes in the DNA sequence itself – take place. These phenomena are explained in Chapter 6. What matters for our discussion here is that the genomes of individuals undergo molecular (genetic and epigenetic) changes during development, and thus are not static. Therefore, the adjective "identical" initially used by Galton and others is not accurate as a description. It is more appropriate to describe these twins as monozygotic twins – that is, twins emerging from the same zygote (fertilized ovum). In this way, we can distinguish them from dizygotic twins – that is, those emerging from two different zygotes that emerged because two ova were fertilized independently. Dizygotic twins are genetically related as much as any two siblings are, and share 50 percent of their DNA on average. Figure 5.7 shows the amount of the common DNA within a family. It must be noted that the percentages therein are averages, with the exception of the parent–offspring comparison because we have all inherited half of our DNA from each of our parents.

In 1969, educational psychologist Arthur Jensen published a paper that suggested that differences in IQ (intelligence quotient) among people were largely due to genetic differences. IQ was expressed as the ratio of a child's mental age to the child's chronological age, times 100. If the ratio was 1, then the IQ would be 100; if the ratio was 1.2, then the IQ would be 120; and so on. This practically meant that a child with an IQ higher than 100 would be more advanced and have a mental age that was ahead of this child's chronological age. In the opposite situation, the child would be considered as somehow having delayed mental development. Jensen's aim was to question the assumption that IQ differences were almost entirely due to differences in the environment in which people were brought up. He compared two extreme cases: monozygotic twins brought up in different families, and unrelated children brought up in the same family. In the first case, the

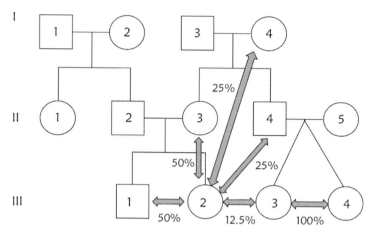

Figure 5.7   Relatives have common DNA. The pedigree illustrates the members of a family (males are represented by rectangles and females by circles; spouses are directly connected with a horizontal line; siblings are connected with a horizontal line on top of them that also connects them to their parents). The extreme case for two siblings is to receive exactly the same 23 chromosomes from each of their parents, which is very unlikely. There are several possible combinations. As they each receive half of their DNA from each of their parents, which in turn corresponds to half of their total DNA, we can estimate that siblings share on average 50 percent of their DNA (e.g., III1 and III2). Only III3 and III4 share (almost) 100 percent of their DNA because they are monozygotic twins. For example, III2 has also 50 percent of her DNA common with her parents (II3 is her mother), and on average 25 percent of her DNA in common with her maternal grandmother (I4) and her maternal uncle (II4). Finally, III2 and III3 are first cousins and have 12.5 percent on average of their DNA in common. Note: all percentages are averages, except for the parent–offspring relations.

available data showed that the correlation between monozygotic twins in different families was 0.75. Jensen argued that as there should be no genetic differences between monozygotic twins, any difference between them should therefore be due to environmental factors. So, Jensen concluded that 75 percent of the variance could be said to be due to genetic variation, and 25 percent due to environmental variation. This was, for Jensen, the definition of heritability. He wrote: "Thus 75 percent of the variance can be said to be due to genetic variation (this is the heritability)."

Jensen then looked at data from unrelated children living in the same family. He argued that as these children were not genetically related, any correlation between them should reflect their common environment. The available data showed that this correlation was 0.24. Therefore, the proportion of IQ variance due to environment was 24 percent and the remaining 76 percent should be due to heredity. Jensen noted the good agreement between these two estimates: 75 percent and 76 percent of differences should be due to genetic factors in these two extreme cases. He also calculated a composite value from all kinship relations available (not just the two extreme cases discussed), and he concluded, after appropriate statistical corrections, that the best single overall estimate for the heritability of intelligence was 0.81. He thus concluded that approximately 80 percent of the variation in intelligence was due to genetic differences. He also concluded that the influence of the environment on IQ was that of a threshold variable: extreme environmental deprivation could keep a child from expressing his/her genetic potential, but no enriched educational program could push a child above that potential. Jensen thus argued that educational attempts to boost IQ had been misdirected.

Perhaps the most influential critic of Jensen's conclusions was Richard Lewontin. In his critique, Lewontin initially accused Jensen of being biased and rushing to pre-reached conclusions rather than questioning the methods with which he reached them. Then he argued that it is not possible to speak of characteristics as being molded by heredity, as opposed to environment. The reason for this, according to Lewontin, is that every characteristic is the outcome of "a unique interaction" between genes and the various environments in which an individual has lived and developed. Sometimes it is genes or environment that are more influential in shaping the developmental outcome, but in all cases the relationship between genes and characteristics, as well as environments and characteristics, is a many-to-many relationship. Lewontin went on to explain why we cannot talk of heritability of a characteristic in general, but only in a particular population and in a particular set of environments, a point to which I return below. Lewontin concluded his critique by making two major points: (1) that in order to explain how genes and environment affect characteristics, it is necessary to understand the respective developmental process; and (2) that conclusions made about the possible contributions of environment may be wrong because of the limited number of different environments studied and because not all possible environments have been considered. In subsequent work, Lewontin drew on data such as those on *Achillea* plants to argue that phenotypes are influenced by differences in both environments and genotypes, and that their various combinations produce a variety of phenotypes. In this sense, both genotypes and environments should be considered as causes of phenotypic differences.

If one reads the exchange between Jensen and Lewontin, one may conclude that there are ideological biases behind the different views and the criticisms. In a thoughtful analysis of this exchange, however, philosopher and historian of science James Tabery has argued that the real problem is what he describes as the explanatory divide. According to Tabery, the debate was about three distinct questions: (1) What is the interaction between nature and nurture? (the conceptual question); (2) why and how should the interaction between nature and nurture be investigated? (the investigative question); and (3) what is the empirical evidence for the interaction between nature and nurture? (the evidential question). According to Tabery, the real reason for the disagreement between Jensen and Lewontin was that they addressed these questions differently. Details notwithstanding, Jensen was interested in explaining the variation in a population by asking how much of this variation is due to one or the other factor and thus by identifying the causes of this variation. In contrast, Lewontin was interested in explaining the developmental process by asking how this takes place and thus by finding the causal mechanism responsible for it. Therefore, their different explanatory aims may have made understanding each another difficult. However, this disagreement also reveals the confusion about the concept of heritability.

Heritability is often defined as the proportion of the variance in a population that is due to genetic variation. Thus, if we say that the heritability of a characteristic in a population is, say, 0.8, this means that 80 percent of the phenotypic variation in this population is due to variation in genetic factors, whereas 20 percent of the observed phenotypic variation can be explained on the basis of the variation in environmental factors. A first point to note is that what heritability is about is the explanation of the variation in a characteristic, and not about the origin of the characteristic itself. However, it is easy to conflate these two. Let us consider an example by developmental biologist Jonathan Slack. Imagine a population of cattle in which the heritability of final size is 50 percent. This means that variation in final sizes is 50 percent due to variation in genes and 50 percent due to variation in environmental factors, both of which are implicated in development. Now imagine that at some point a farmer buys all the farms in the area. If the farmer started artificially selecting the same type of cattle, with a very specific final size, after several generations genetic variation would probably be smaller compared to the beginning, because some variants that were not selected would no longer be present in the population. If environmental variation had been kept constant in the meantime (i.e., the differences in nutrition and husbandry among the farms continued to exist during that time), eventually the heritability of final size would fall because of the reduction in genetic variation. In contrast, if the farmer instituted the same conditions in nutrition and husbandry in all farms, and thus the environmental variation was reduced, the heritability of the final size would increase. This would be possible, despite no changes in the genetic variation, exactly because the environmental variation was reduced. What this example shows is that heritability is context-dependent and any measurement is valid only for a particular population under particular conditions. This is why we cannot predict the heritability of a characteristic, even when we know that it is genetic, and we cannot predict if a characteristic is genetic, even when it is known to have high heritability.

A second important point is that heritability is not about measuring the relative, additive contribution of genes and environment to phenotypes. This is impossible to measure because these contributions are interdependent, not independent. Here is a way to illustrate this, suggested by Lewontin. Imagine two men who lay bricks to build a wall. In this case, it is quite easy to measure the contribution of each of them to the final outcome. If man A laid 40 bricks and man B laid 60 bricks to build the wall, we can say that the building of the wall was 40 percent due to man A and 60 percent due to man B. In this case, the causal influence of the two men is independent and thus measurable. However, if man A mixes the mortar and man B lays the bricks, it is not possible to measure their relative contributions. Even if we could measure the quantity of mortar and bricks in the wall, this would not be a measure of each man's contribution as both mortar and bricks are needed for building the wall and cannot be considered separately. In this case, it is impossible to distinguish between the causal influences of the two men, because they interact and depend on each other to build the wall. In this sense, the question about the relative contribution of genes and environment to the production of a characteristic makes no sense, because they do not make independently measurable contributions.

However, even experts often make confusing and misleading claims as if it were possible to clearly distinguish between genetic and environmental influences. Why such claims are wrong is best illustrated by an example provided by developmental psychologist David Moore: the formation of snowflakes. This phenomenon requires the simultaneous presence of two factors: temperatures below 0 °C and a relative humidity sufficient for precipitation. Imagine now that on a given day humidity is high at the North Pole but low at the South Pole, where the temperatures are below 0 °C anyway, and that as a result there is formation of snowflakes and snowfall only in the North Pole. In this case, the variation in snowfall across the two places can be entirely accounted for by variation in relative humidity. However, this does not entail that it is relative humidity alone that caused the snowfall in the North Pole; the low temperature is also necessary for this. Therefore, accounting for variation is very different from causally explaining a phenomenon. Differences in humidity alone are sufficient to account for differences in snowfall between the two poles only because the temperatures there are always below 0 °C and not because they are unimportant in causing snow. I hope this example shows why heritability estimates can only provide us with information about what causes variation in characteristics, and not at all about what causes the characteristics themselves.

These examples should make clear why high heritability does not indicate that a characteristic is more influenced by genetic factors than by environmental factors. In fact, even strongly inherited characteristics can have heritability equal to zero in particular populations. For example, humans normally have two legs and two arms. Differences in DNA that cause differences in the number of limbs in humans are rare, and so most differences are explained in terms of differences in the environment (e.g., accidents leading to amputation, environmental influences during development). As a result, whereas our

DNA indicates that we should have two legs and two arms, instead of one or three, these characteristics have a very low heritability. Therefore, heritability estimates tell us nothing about what causes one's characteristics.

Genes Account for Variation in Characteristics in Populations

Genes cannot account for the relative contribution of genes and environment to a characteristic, but they can account for characteristic-differences in a population. Let us consider an example to illustrate this, suggested by Evelyn Fox Keller. Imagine a drummer playing his drums. It makes no sense to ask whether the drumming sound we hear is mostly due to the drummer's competence or to the quality of the instrument he is playing. The reason for this is that the outcome depends on both. However, what makes sense to compare are two different drumming sounds, and to explain not the drumming sounds themselves but the differences between them. In the case of two drummers playing the same instrument, any difference between the drumming sounds would be almost 100 percent due to the drummers. In the case of the same drummer playing different instruments, any difference in the drumming sound would be almost 100 percent due to the instrument. In each of these cases, there is a single difference-maker, the drummer and the drums, respectively. Now, the situation becomes more complicated in the case of two different drummers playing two different instruments. In this case, the difference will be somewhere in between the two previous cases, depending on the competence of the drummers and on the quality of the instruments. The crucial point is that what can be explained on the basis of a single gene is not a certain characteristic (e.g., why an individual has blue eyes or why another has brown eyes), but differences in characteristics (e.g., why more Scandinavian people than Mediterranean people have blue eyes).

As already discussed in Chapter 2, it was clear to Morgan and his collaborators that several genes (at the time still called factors) were necessary to produce a characteristic, but also that a change in one of them was sufficient to bring about a change in that characteristic. This means that although single-gene variation may explain characteristic-differences in a population, the development of a characteristic is not sufficiently explained by the activity of a single gene. It is in this sense that genes are difference-makers for their effects. Thus, for instance, *SRY* is a difference-maker for the development of sex, but not the "gene for" sex determination. This means that a change in the presence of *SRY* (depending on whether a human has a Y chromosome or not), or in its expression (depending on whether the *SRY* gene is expressed or not due to a mutation) can make a difference in the development of testes or ovaries. But the *SRY* gene alone cannot account for the development of these organs and sex (as mentioned in the previous chapter, a mutation that would affect SOX9 would impact the development of these organs too).

To better understand what a difference-maker is, let us consider a classic example. A forest fire can be the effect of several different causes, such as a lighted match, oxygen, and flammable materials such as dried grass. There will be no forest fire if there is no oxygen because it is necessary for combustion; flammable material such as dried grass is also a cause, because wet or humid material is difficult to burn. However, in the case of a forest fire, it is a lighted match that is considered to have caused the fire. Nobody usually considers the presence of oxygen or the dried grass as the main causes of the fire, although there is no question that these are causes too. The reason for this is that oxygen and dried grass exist in a forest all the time (at least during summer), but the initiation of a fire is not due to any change in them. In contrast, it is a change in the condition of a match that makes the difference and initiates the fire. Therefore, in the case of a forest fire, there are at least three causes (match, oxygen, dried grass), but it is the lighted match that is the cause that makes the difference – it is the difference-maker. More generally, whereas there can be several different causes for a particular effect, there is only one or a few that are more important than the others – causally speaking – because it is their differentiation that brings about the effect. It is in this sense that genes have an explanatory power and an enormous heuristic value in genetics research. An effect can be explained if one identifies the actual difference-maker – that is, the cause that explains why some variable varies in a population, such as the lighted match in the forest fire example.

It should be noted that the concept of difference-makers applies only to causes in populations with members that actually differ with respect to some variable that is the effect of these difference-makers. It should also be noted that such an effect is not the property of a single individual, but a difference of a property within a population. Therefore, for example, the population of Greece consists of some people exhibiting symptoms of β-thalassemia and even more people without any such symptoms. In this case, the difference-makers are the mutated *HBB* alleles that were actually found in the particular β-thalassemia patients. This means that phenotypic differences between healthy people and β-thalassemia patients are due to differences in the *HBB* alleles, or that variation in the phenotype with respect to β-thalassemia is due to variation in the *HBB* alleles. Therefore, we can say that differences in the *HBB* alleles cause differences in the phenotype between healthy people and β-thalassemia patients; however, this is not equivalent to saying that the *HBB* allele causes thalassemia. The latter should be interpreted as the former, but by stating the former one does not also state the latter.

We should also keep in mind that genes are not the only difference-makers, even for diseases that are considered to be genetic. For example, lactose intolerance in humans (the decreased ability to digest lactose, a sugar found in dairy products) can be viewed either as a genetic disease or as an environmental disease depending on the population considered. The reason for this is that individuals in different populations could be healthy for different reasons, depending on whether or not they have the allele related to lactose intolerance and on whether or not they consume lactose. As a result, lactose intolerance is considered to be a genetic disease in populations in which ingestion of milk products is common and lactase deficiency is rare, because it is the latter that actually makes the difference in the occurrence of the disease (most people do not have the respective allele, and so they have no problem even though they consume milk products). In contrast, lactose intolerance is considered to be an environmental disease in populations in which ingestion of milk products is rare and lactase deficiency is common, because it is the former that makes the difference in the occurrence of the disease (most people do not consume milk products, so they have no problem even though they have the respective allele). Therefore, a characteristic cannot be described as "genetic" in any absolute sense, as this depends on the respective population. This is important because either the presence of an allele or the presence of an environmental factor could make the difference in the occurrence of a characteristic or a disease. What matters is not only whether a gene is associated with a characteristic or a disease, but also the detailed understanding of why and how changes in a gene might bring about a certain version of a characteristic or a disease.

The important conclusion of this chapter is that developmental conditions and influences affect why, when, and how our genome is expressed. In other words, our "nurture" affects our "nature"; this does not make the latter less important than the former. Actually, it is only accurate to say that the two interact. Thus, genes do not act alone, but with their environment, and it is this interaction that drives the development of our characteristics. It is perhaps time to abandon altogether the notion of "gene action," and always think in terms of "gene interaction." This is why single genes can only account for characteristic-differences and not characteristics themselves. Genes do not invariably determine characteristics and disease, but characteristics are affected by changes in genes. Characteristics are also affected by changes in the environment or by the different environments in which an individual lives. Actually, in some cases the cellular and the organismal environment impact the expression of genes, as I explain in the next chapter.

To summarize, what is the answer to the question "what do genes do?" First, genes do nothing on their own; they must be activated or inactivated by particular molecules (mostly proteins) in order to produce the RNA or protein molecule that in turn is implicated in some process that results in a characteristic or disease. Second, as explained in Chapter 3, genes cannot even be isolated and analyzed; it is DNA that can be isolated and analyzed. Yet, even DNA cannot do anything on its own. It is more appropriate to think of DNA as the primary resource on which cells draw. But even in that case, DNA is part of a complex system of interacting molecules that altogether guide the development of characteristics. Therefore, if we would like to summarize the several, often overlapping, roles of DNA within cells, and answer the question "What do genes do?" we might conclude that: (1) genes are resources for the production of functional molecules (proteins or RNA), which are implicated in development and physiology; and (2) they account for variation of characteristics in populations because differences in characteristics can be due to differences in genes.

6 The Dethronement of Genes

Genes Are Not Master Molecules: The Relation Between Genes and Characteristics or Disease Is Complex

One important, and for some the most surprising, conclusion of genome-wide association studies (GWAS) has been that in most cases numerous single nucleotide polymorphism (SNPs) in several genes were found to be associated with the development of a characteristic or the risk of developing a disease. As already mentioned, the main conclusion has been that the relationship between genes and characteristics or diseases is usually a many-to-many one, as many genes may be implicated in the same condition, and the same gene may be implicated in several different conditions. In fact, the same allele may be protective for one disease but increase the risk for another. For example, a variation in the *PTPN22* (protein tyrosine phosphatase, non-receptor type 22) gene on chromosome 1 seems to protect against Crohn's disease but to predispose to autoimmune diseases. In other cases, certain variants are associated with more than one disease, such as the *JAZF1* (JAZF1 zinc finger 1) gene on chromosome 7 that is implicated in prostate cancer and in type 2 diabetes. Therefore, we should forget the simple scheme of gene 1 → condition 1/gene 2 → condition 2, and adopt a richer – and certainly more complicated – representation of the relationship between genes and disease. Additional GWAS on more variants in larger populations might provide a better picture in the future. But insofar as we do not understand all biological processes in detail, all we are left with are probabilistic associations between genes and characteristics (or diseases). The "associated gene" may be informative, but its explanatory potential and clinical value are limited – at least for now.

In order to describe the phenotypic variation observed in the various studies, new concepts such as penetrance and expressivity were invented. Penetrance is the proportion of individuals with a given genotype who have a typical associated phenotype. For β-thalassemia, penetrance could be close to 100 percent as an individual with two defective *HBB* alleles will most likely have the disease. However, in other cases such as breast cancer the penetrance of the *BRCA1* alleles is, on average, 50 percent at the age of 50 and 85 percent at the age of 70. In other words, penetrance is the concept indicating that having an allele related to a disease does not necessarily entail having the disease; it also describes how likely this is. Expressivity indicates the qualitatively different phenotypes that individuals with the same genotype can exhibit, or in other words, the degree to which a particular phenotype appears given a certain genotype. For example, some people with a genotype associated with a disease (e.g., with the same *LDLR* alleles) may experience mild symptoms because they also have a "protective" *PCSK9* allele, whereas others may experience severe symptoms because they do not have a "protective" allele. The phenotype of an individual may indicate a genotype; however, the opposite does not work as we cannot always predict a phenotype from a genotype. The reason for this is that it is developmental processes, not genes, that actually produce the phenotype.

That inferences about the presence or absence of a certain gene from the phenotype may be wrong has been most elegantly demonstrated by the process of "gene knockout," the targeted disruption of particular genes in their actual biological context. Gene knockout involves the replacement of the "normal" copy of a gene with an abnormal copy, which in turn results in the production of a nonfunctional protein or no protein at all. This is expected to produce a phenotype lacking some function that could then be attributed to the knocked-out gene, as its disruption would be the most plausible explanation for the loss of the particular function. However, even in the cases of genes that were considered as essential for some function, the knockout procedure has not always given the anticipated effects, and has thus shown, perhaps in one of the clearest ways, that there are no "genes for" characteristics. A very likely reason is that many genes are implicated in a certain function, and if one of them is not functioning the others may compensate for it.

Gene knockout technology is the combination of two techniques: the culture of multipotent embryonic stem cells from mouse embryos, and the introduction of mutations into these cells by homologous recombination. This is a procedure used to insert an allele to a specific homologous genetic locus. It can be performed in embryonic stem cells, which are then injected into an early embryo (blastocyst) and differentiate in all types of cells. As a result, this embryo will give rise to a chimeric organism – that is, one that has two types of cells: the "normal" ones and those with the knockout gene. This chimeric organism can later breed and give rise to offspring that will have the knockout gene in all their cells.

Studies with knockout mice have revealed different kinds of phenomena. (1) Many genes encode related molecules and so it is difficult to define their precise function due to their overlap with others. (2) Knockouts of the same gene with different methods can result in different phenotypes. (3) Even when the same knockout method is used, differences in the genotypes of mice can produce differences in phenotypes. (4) The disruption of a gene might affect the expression of another one close to it, thus producing an altered phenotype that is not connected to the gene of interest. (5) The same mutation in different conditions can give rise to different degrees of disease occurrence. (6) Disruptions in genes that affect an organ, such as the lymph nodes or the spleen, might have secondary effects on the function or differentiation of its cells, which might be misinterpreted as primary effects of the gene disruption on the cells themselves. These observations once more indicate clearly that the relation between genes and phenotypes is a many-to-many one.

This kind of research with knockout mice has improved our understanding of human disease. Yet, from the results in mice we can only make inferences about the respective human conditions, as there are important differences between mice and humans. At the same time, experimentation with humans is, of course, not an option. However, there exist "natural knockouts" among humans – that is, people with mutations that have resulted in the loss of function for particular genes – the study of whom has yielded some surprising conclusions further indicating the complexity of the relation between genes and characteristics. Of course, such mutations have resulted in disease, including those discussed in the previous chapter. However, modern sequencing methods that span the whole genome have also allowed the identification of human knockouts without any accompanying disease phenotype. An interesting observation has been that even for genes that have a well-established connection to disease, natural human knockouts have very different phenotypes than those of individuals having only one functional allele of the respective gene (this phenomenon is called haploinsufficiency). For example, carrying a certain *BRCA2* allele is associated with a risk for breast cancer. However, human knockouts for *BRCA2* have different phenotypes such as primordial dwarfism, a disease with severely impaired fetal growth that results in shorter size in adulthood. Interestingly, it has been estimated that human genomes contain approximately 100 genuine loss-of-function variants with approximately 20 genes completely inactivated. It seems that this phenomenon is quite common and varies depending on the organ or tissue involved as, for example, genes that are highly expressed in the brain are less often completely knocked out than other genes. Another study that focused on healthy individuals aimed at identifying whether there could be healthy carriers of gene variants associated with highly penetrant forms of disease. The comprehensive screening of 874 genes in 589,306 genomes led to the identification of 13 adults who had mutations for 8 severe Mendelian conditions, but no clinical symptoms of the respective diseases.

Another important phenomenon, also revealed by GWAS, is pleiotropy. In the previous chapter I mentioned genes that may be associated with, and thus assumed to be implicated in, more than one disease. More broadly, there exist associations between gene variants and more than one characteristic, and these are described as cross-phenotype associations. However, in these cases we do not know the underlying cause of the observed association. Pleiotropy occurs when a genetic variant *actually* affects more than one characteristic, and it is thus a possible underlying cause for an observed cross-phenotype association. In other words, pleiotropy can be identified only when we have understood the underlying mechanism. Overall, we can distinguish between three different types of pleiotropy. The first is biological pleiotropy, which occurs when a genetic variant or gene has a direct biological effect on more than one phenotype. For example, the common coding variant in *PTPN22* (protein tyrosine phosphatase, nonreceptor type 22) on chromosome 1, which has been found to be associated with disorders related to the immune system (such as rheumatoid arthritis, Crohn's disease, systemic lupus erythematosus, and type 1 diabetes), seems to influence the function of various types of T-cell and to also be involved in the removal of B-cells. A second type is mediated pleiotropy, which occurs when one phenotype is itself causally related to a second phenotype, so that a genetic variant or a gene associated with the first phenotype is indirectly associated with the second phenotype. For example, in the previous chapter I mentioned that several genetic variants have been found to be associated with both low-density lipoprotein (LDL) levels and the risk of coronary heart disease. However, LDL levels are themselves risk factors for coronary heart disease. Therefore, what must be clarified is whether a genetic variant influences coronary heart disease risk by altering LDL levels or whether it has another effect that is independent of LDL levels. Finally, spurious pleiotropy occurs when a genetic variant falsely appears to be associated with multiple phenotypes, but it is actually not related. This is usually due to faults in the study design, such as collecting a nonrandom subsample with a systematic bias so that results based on it are not representative of the entire sample (ascertainment bias), or when individuals with one phenotype are systematically misclassified as having a different phenotype.

To put it simply: The relation between genes and characteristics is complex. Therefore, genes are neither master molecules that determine phenotypes, nor is identifying the impact of individual genes on phenotypes sufficient for successful medical interventions.

Genomes Are More than the Sum of Genes: Epigenetics

As described in Chapter 3, DNA is not "naked," but packed with proteins forming chromatin. One important finding of the ENCODE project was that the state of chromatin and transcription are related, although there is still a lot to figure out. There are several phenomena producing modifications in chromatin. These cause no actual change in the nucleotide sequence of genes, but they can strongly affect gene expression – that is, whether or not a protein or an RNA molecule is produced. These modifications can be permanent and may even be transferred across generations. As this is inheritance of modifications that are not made in the DNA sequence but "upon" it, it is described as epigenetic inheritance (the prefix "epi-" literally means "upon" in Greek). The term epigenetics currently refers to all interactions between DNA and its local environment that eventually influence gene expression. Let me get back to *The Phantom of the Opera*. Imagine that a play director took the script of *The Phantom of the Opera* and made modifications. Some of these modifications can be permanent if the play director made notes on the script with a pen. But some other modifications can be reverted, introduced again, and then reverted again, and so on, as if a play director took the script, made notes with a pencil, erased them, made the same or new notes again, and so on. These modifications affect whether genes are activated or deactivated, and therefore expressed or not expressed, under different conditions.

Let us see what kind of modifications take place. The first of these modifications is called DNA methylation. Simply put, methylation is the addition of a small chemical group called methyl ($-CH_3$) to cytosine (C). This is done by enzymes called DNA methyltransferases (DNMTs – as their name indicates, they transfer methyl groups to DNA). Thus, a molecule called methyl-cytosine, or methyl-C, is produced. More specifically, the methyl group is added to a C that is followed by a G in the same DNA strand. The sequences consisting of several pairs of these two consecutive bases are described as CpG islands. These are not randomly distributed in the genome but are usually found upstream (before) genes, within their promoters. When CpG islands are highly methylated, the respective genes are not expressed; in contrast, genes with a low level of methylation in CpG islands are expressed. The protein MECP2 (methyl-CpG binding protein 2) binds to highly methylated DNA and blocks the expression of the respective gene by attracting other proteins that also bind there or by not allowing RNA polymerase to produce mRNA. What is even more important is that the methylation pattern of a DNA molecule is preserved after DNA replication that precedes cell division, and thus the new cells inherit it. This entails that whatever DNA sequence an individual has in a given gene, it may not make much of a difference if this gene is not expressed. The important idea here is that the script of *The Phantom of the Opera* does not only include information about who says what, but also about when one says anything and even whether one speaks at all. The director may use a pen to cross out some lines in the initial script, to simply put them within parentheses, or to make some notes. In the case of DNA methylation, these changes can be permanent. Whatever the message in these lines, it makes no difference if these are crossed out and nobody reads them.

There are other types of changes that are less permanent, as if the play director used a pencil – not a pen – to cross out lines, put them within parentheses, or write notes. These are changes in the histone molecules, which have an impact on gene expression. The first change of this kind to be identified was histone acetylation, which is the addition of an acetyl group (CH_3CO-) to the amino acid lysine in particular histones. This seemed to facilitate the binding of DNA transcription factors and thus to result in increased expression of nearby genes. Several other changes like this, which either disrupt chromatin or affect the binding of nonhistone proteins to it, can affect the structure of chromatin and thus whether enzymes can access DNA. One example is histone methylation, which is the addition of a methyl group ($-CH_3$) to a specific lysine of a particular histone. What is even more interesting is that histone modification and DNA methylation seem to be related phenomena, as the former seems to affect the occurrence of the latter. For example, in mammals, DNA methylation and methylation of lysine at position 9 of histone H3 are strongly associated, whereas methylation of lysine 4 of histone H3 seems to inhibit DNA methylation. Therefore, histones are more than just a means of packaging DNA, as histone modifications can affect DNA methylation and in turn gene expression. Which epigenetic marks (DNA methylation, histone methylation, or histone acetylation) a cell contains (Figure 6.1), in other words the overall epigenetic state of a cell, is described as its epigenome. These modifications seem to form a special, complex code that is "read" by proteins, and that in turn affects the accessibility and function of eukaryotic genomes.

The effect of epigenetic modifications is easy to observe in calico cats, which have a coat with a color that is a mixture of black and ginger spots. You can impress your friends by guessing that these cats are females just by looking at their coat color. What happens in this case is that the gene that affects coat color is found on X chromosomes. A female cat can be heterozygous and carry an allele that is implicated in the development of black color (B) and

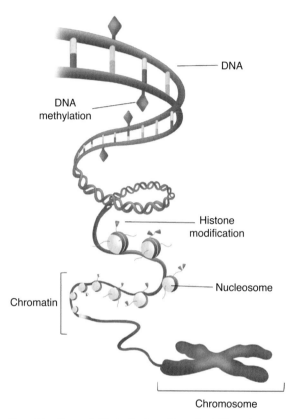

Figure 6.1   Two main kinds of epigenetic marks: DNA methylation and histone modification. In DNA methylation, methyl groups are added to certain cytosines. In this way, gene expression is repressed. In histone modification, a combination of different molecules can attach to the "tails" of histones. These alter the activity of the DNA wrapped around them (© Media for Medical).

another allele that is implicated in the development of ginger color (G); therefore, it can have the genotype X^BX^G. However, in female cats one of the X chromosomes becomes inactivated. Thus, both males and females actually have a single, active X chromosome in each cell. In female calico cats, it is the X^B chromosome that is inactivated in some cells, whereas in other cells it is the X^G chromosome that is inactivated. As a result, the coat of the cat becomes a mosaic of colors because it is a mosaic of cells with different inactivated X chromosomes, expressing either only the B or the G allele.

A relevant but different phenomenon is that of genomic, or epigenetic, imprinting. In some cases, the chromosomes of one of the 23 pairs that each of us has in our cells cannot be distinguished from each other. There is nothing that indicates which of the two homologous chromosomes was inherited from one's mother and which was inherited from one's father. However, for some chromosomes, such as chromosome 11, this is possible. Mammals have evolved to have epigenetic marks that distinguish between the paternal and the maternal chromosomes. But this is not just a distinction issue; maternal and paternal chromosomes are not always equivalent. In the early 1980s, nuclear transplantation experiments showed that both parental genomes are necessary for complete development. An interesting finding emerged in experiments with mice that had inherited one chromosome of each pair from each of their parents, except for chromosome 11. In that case, some mice had inherited either two paternal chromosomes 11, or two maternal chromosomes 11 (this condition is described as uniparental disomy). The observed outcome was a consistent difference in the size of these mice. Mice having both a paternal and a maternal chromosome 11 had regular size, but mice with chromosomes 11 of maternal origin were smaller than the others, and mice with chromosomes 11 of paternal origin were larger than the others (this was the case for both male and female mice).

These results showed that maternal and paternal chromosomes are not functionally equivalent. But why is this happening? It seems that certain genes that are switched on in one chromosome are switched off in the other chromosome. Therefore, when an organism has two chromosomes of the same kind, there is hyperactivity or hypoactivity of these genes. In the case of the mice described in the previous paragraph, genes related to placental growth were very active in the paternally derived chromosomes and not very active in the maternally derived chromosomes. This "switching on" and "switching off" of particular genes of the same chromosome is due to epigenetic marks that exist on the chromosome of the one parent but not on that of the other. This phenomenon is described as genomic imprinting. During this process, the alleles of particular genes acquire an epigenetic mark (e.g., DNA methylation) at specific sequences called imprinting control regions. As a result, only one of the alleles of these genes is expressed in the offspring, and these are distinguishable in terms of parental or maternal origin. This imprinted gene expression is maintained by various mechanisms such as noncoding RNAs and histone modifications. Recent research has shown that genomic imprinting varies between genes, individuals, and tissues. An interesting instantiation of this has been observed for a long time. It has been long known that mating a female horse and a male donkey produces a mule, whereas mating a male horse and a female donkey produces a hinny. Hinnies are a bit smaller than mules, and they also differ in their physiology and temperament. It seems that these differences are due to imprinting.

The most well-known examples of genomic imprinting in humans are Prader–Willi and Angelman syndromes. In both cases, the syndromes are associated with the deletion of the same part of chromosome 15, and what makes the difference is whether the chromosome with the deletion was inherited from the mother or from the father. In particular, Prader–Willi syndrome is associated with the deletion on the paternal chromosome 15, whereas Angelman syndrome is associated with the deletion on the maternal chromosome 15. In people with Prader–Willi syndrome, the paternal chromosome has the deletion, and the respective section on the maternal chromosome is imprinted and thus unexpressed. However, it is not the deletion that causes the problem, as people with Prader–Willi syndrome have been found to have two complete chromosomes 15. It seems that these are two maternal chromosomes that are imprinted and thus non-expressed. In the case of Angelman syndrome, the deletion exists on the maternal chromosome 15, and it is the paternal chromosome 15 that is imprinted and unexpressed. From these observations, it becomes clear why these syndromes are both genetic and epigenetic in nature. The effects of a chromosome deletion are not compensated by the respective sequences on the other chromosome because the latter are imprinted and thus switched off. Furthermore, where this happens makes a difference to which syndrome these people will develop. This is because the genes involved in each case are not exactly the same. Prader–Willi syndrome is due to loss of paternal expression of up to 11 genes as a result of the deletion in the paternal chromosome. Angelman syndrome arises from loss of maternal expression of the *UBE3A* (ubiquitin protein ligase E3A) gene, mainly due to the deletion in the maternal chromosome.

An even more interesting finding is that epigenetic differences can account for phenotypic differences between monozygotic twins, who are otherwise supposed to have (almost) the same DNA sequences. A study with 80 monozygotic twins showed that although they were indistinguishable during their early years of life in terms of DNA methylation and histone acetylation, in older twins there were remarkable differences in the genomic distribution of these marks that affected gene expression. These differences were more significant between older monozygotic twins who had different lifestyles, and had spent less of their lives together, thus highlighting the significant impact of environmental factors on the development of different phenotypes on the basis of the same genotype. Interestingly, although the differences are small, there is variation in DNA methylation even between newborn monozygotic twins, as well as between different tissues. Such results suggest that the intrauterine period is a crucial period for the establishment of epigenetic marks in humans. This means that no matter what DNA or gene sequences one has, knowing it is not enough for predicting the phenotype. Which epigenetic marks one has and why also makes a crucial difference.

A striking example of how epigenetic changes can impact adult characteristics is that of the so-called Dutch famine. During the winter of 1944–1945, while the Netherlands was under German occupation, there were shortages of food in the western part of the country due to a German-imposed food embargo. Although the southern part of the country was already liberated by the Allies, this was not yet the case for the northern part. To support the Allies, the exiled Dutch government organized a strike of the national railways in an attempt to delay the transfer of German soldiers. The Germans responded by putting an embargo on all food transports. Even though the embargo was partially lifted in November 1944 by allowing transportation of food across water canals, an unusually severe winter caused all canals to freeze and so food transportation became impossible. As a result, the food stocks diminished. Between December 1944 and April 1945, people in the affected region, including pregnant women, received around 400–800 calories per day, and were thus seriously undernourished. A study looked at 2,414 people born between November 1943 and February 1947 in Amsterdam, for whom detailed birth records were available. Those people whose mothers had received less than 1,000 calories per day during the first, second, or third trimester were considered exposed to the famine, whereas those who never received less than 1,000 calories per day were considered unexposed to the famine. The researchers found that the long-term consequences of exposure to famine varied depending on the trimester of the gestation during which the exposure to famine occurred. People born to mothers exposed to famine in the first trimester of pregnancy were found to have more coronary heart disease, a more atherogenic lipid profile, disturbed blood coagulation, increased stress responsiveness, more obesity, and – for women – increased risk of breast cancer. People born to mothers who were exposed to famine in the second trimester had more microalbuminuria and obstructive airways disease. Finally, all people born to mothers who were exposed to the famine at any stage had reduced glucose tolerance and increased insulin levels. But what does epigenetics have to do with these outcomes? Well, it seems that epigenetic differences underlie these different outcomes.

A study of people affected by the Dutch famine aimed at investigating whether prenatal exposure to that famine is associated with differences in methylation of the *IGF2* gene (insulin like growth factor 2, on chromosome 11). For this purpose, 60 individuals who were conceived during the famine and 62 individuals who were born right after it were selected to participate in the study. The levels of *IGF2* DMR (differentially methylated region, located upstream of the gene) methylation were then compared between these individuals and their siblings. In the first group (exposed to famine during early pregnancy), *IGF2* DMR methylation was found to be lower than for their siblings for 43 out of 60 individuals. In contrast, no difference was found between the individuals of the second group and their siblings. It was concluded that those people who were exposed to the famine around the time of their conception and early in their mother's pregnancy had overall lower methylation of the *IGF2* DMR 60 years later. In contrast, those people who were exposed to famine during the later stages of pregnancy did not have this feature. This means that their mothers' exposure to the famine during early pregnancy had long-lasting implications for the epigenome of these people.

Epigenetic marks are so important that they can actually determine the potency of a cell, which can be one of the following:

1. Totipotent – that is, divide and eventually produce all the cell types of the body and the placenta. The blastomeres of the early embryo are totipotent cells.
2. Pluripotent – that is, divide and eventually produce all the cell types of the body but not the placenta. The cells of the inner cell mass of the blastocyst are pluripotent cells.
3. Multipotent – that is, divide and eventually produce various cell types of the body. Most adult stem cells, such as skin stem cells, hematopoietic stem cells, and neural stem cells are multipotent.
4. Unipotent – that is, divide and eventually produce only one cell type. Examples of such cells are differentiated cells, certain adult stem cells (e.g., testis stem cells), and certain progenitor cells (e.g., erythroblasts).

What is now clear is not only that the cells of various kinds of potency are in different epigenetic states, but also that differentiated cells can revert to less differentiated states if their epigenetic state changes. Adding particular transcription factors can turn somatic cells to a pluripotent state without the use of oocytes, generating what have been called induced pluripotent stem (iPS) cells. It is important to note at this point that whereas epigenetic modifications are generally stable in the somatic cells of mammals, all or most (this is not entirely clear yet) of the epigenetic marks (DNA methylation, histone acetylation, histone methylation, etc.) are erased and are created anew on a genome-wide scale at two distinct stages: in the primordial germ cells (PGCs) when they have reached the embryonic gonads, and in the early embryo between the stage immediately after fertilization and the blastocyst stage. This process is described as epigenetic reprogramming, and it contributes to resetting the epigenetic marks. This is important as, for example, erasure of the epigenetic marks after fertilization is necessary in order for the cells of the early embryo to acquire pluripotency.

Although epigenetic phenomena do not entail the transfer of any information, they affect the expression of the information encoded in DNA. This makes necessary a reconceptualization of the central dogma (Figure 6.2) that also takes into account the effect of transcription factors and

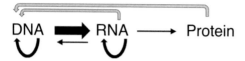

Figure 6.2   The postgenomic "central dogma." The gray arrows do not represent any flow of information, like the black ones do, but rather influences that affect the flow of information represented by the black arrows. The relative sizes of arrows are also indicative that the flow of information from DNA to RNA is larger than all the others, as active RNA transcripts are more numerous than proteins. This figure encompasses the complexity of DNA expression better than the typical representation found in textbooks, as it also shows the important effect of transcription factors and noncoding RNAs.

epigenetic phenomena. Actually, this should not be called a dogma at all, exactly because it can undergo changes, as with all scientific knowledge, whereas dogmas are not supposed to change. Nevertheless, for historical reasons we can retain the term "central dogma," and based on the research presented in this chapter propose its postgenomic version as in Figure 6.2. In my view, this highlights the importance of recent findings of genomics research. What is certainly necessary to show is the influence of proteins such as transcription factors and of noncoding RNA in the expression of genes. Genes are not autonomous entities but are subject to control and regulation of expression that depends on environmental signals, as well as numerous proteins and RNA molecules.

To put it simply: There is a lot in our genome and on our chromosomes, outside genes, that affects the development of characteristics and disease.

Association Does Not Equal Causation: Genetic Tests and the "Associated" Gene

Given all this, the question now becomes: Is knowledge about genes useful for medical purposes? Do genes really predispose us to certain diseases? Is it helpful to have a genetic test in order to know one's alleles related to some disease in order to take some precautionary or therapeutic action? Francis Collins, director of the National Institutes of Health (NIH), has suggested that knowledge about our genes and DNA can be really useful, notwithstanding the limitations of the findings of GWAS:

> We are on the leading edge of a true revolution in medicine, one that promises to transform the traditional "one size fits all" approach into a much more powerful strategy that considers each individual as unique and as having special characteristics that should guide an approach to staying healthy. Although the scientific details to back up these broad claims are still evolving, the outline of a dramatic paradigm shift is coming into focus.

Collins certainly acknowledges that we are not there yet, but we have entered the era of personalized, genomic medicine. He is of course right that people respond differently to diseases and therapies, and that therefore knowing the physiological peculiarities of each individual is important. Collins' conclusion was that "Your DNA helix, your language of life, can also be your textbook of medicine. Learn to read it. Learn to celebrate it. It could save your life."

Personalized medicine is unquestionably important. What is questionable, however, is whether knowledge about genes and genomes currently makes any difference. Genomic medicine, to which Collins refers, and personalized medicine are not synonyms. Genomic medicine is about the impact that one's genes have on disease. Personalized medicine is about taking individual characteristics into account, and these are not restricted to one's genomic information; these also include one's lifestyle, the environment in which one lives, and a lot more. Therefore, treating the terms as synonymous is misleading because it is like implicitly accepting that our characteristics can be reduced to our genes.

According to Eric Lander, who we met earlier, there exist "3600 genes for rare mendelian disorders, 4000 genetic loci related to common diseases, and several hundred genes that drive cancer." How many of these do you really want to be tested for? How much knowledge of this kind is enough? Worse than that, with these numbers it is very likely that each one of us will carry several variants associated with a disease. In that case, how can we decide whether one is "healthy" or a "carrier," and the limit between the two conditions? To make such decisions, one should know which of these genes or loci are actually doing harm and which are harmless. This is presently very unclear for many of these. Lander suggests that regulation is necessary, pertaining to two very important aspects: analytical validity and clinical validity. For gene-based tests these aspects depend on the answers to the following two questions: "Does the test accurately read out a targeted set of DNA bases in the human genome? Does the targeted set of DNA bases provide meaningful clinical information?" Lander wrote these lines as a response to an article coauthored by Francis Collins that presented initial ideas about the implementation of the Precision Medicine Initiative announced by President Obama in 2015. Assuming that the various studies are successfully replicated and that the companies providing genetic tests clearly show what they are measuring and how, it could be possible to address the first question. The second question, which directly affects every interested citizen, is more difficult to address. As I have explained, the "associated" gene does not provide us with much information. Finding an association between a DNA variant and a disease does not necessarily entail that someone having the variant will develop the disease. There are various reasons for this, including that different genes contribute to one's overall risk of developing a disease, with each one making only a minor contribution; that interactions among genes produce different combinations with different phenotypic outcomes, and so diseases are heterogeneous; and finally that the interaction between genes and environment is always crucial, especially in the case of complex disease. Therefore, knowing one's genes (or DNA variants) associated with a disease is just one piece, or a few pieces, of the puzzle.

Until researchers are able to understand a biological process and how a change in one or more DNA sequences affects a characteristic or the onset of a disease, all we can have are statistical associations. This became evident several years ago, as soon as the first results of direct-to-consumer genetic tests (DTCG tests) became widely known. Several people such as journalists Nic Fleming and Boonsri Dickinson, as well as genetics experts such as Francis Collins and Kevin Davies, who was the founding editor of the journal *Nature Genetics*, had their DNA analyzed in the early days of DTCG testing, each one of them by three different companies (Fleming by GeneticHealth, deCODEme, and 23andMe, and all the others by

Navigenics, deCODEme, and 23andMe). The results and the estimated risk that each of them received from the three companies were not always the same, and in some cases they were even contradictory. Around the same time, the results of 23andMe and Navigenics on 13 diseases for five individuals were compared, and the same conclusion was reached: The different companies provided individuals with different risks for the same disease.

How can the different estimates be explained? This largely depends on the set of variants that each company tests for to calculate the relative risk for each individual. As already explained, the findings of GWAS point to associations between certain variants and certain diseases. However, the various companies can be rather selective in which variants they analyze for a certain disease. This results in contradictory estimations of risk in developing a certain disease. Differences can also occur in the estimations of the same company across time, if over time new associated variants are found and are included in the tests. Thus, the estimated probabilities for each disease for each individual might be different today than in 2009, given the currently available data and that available in 2009. But this is not the most important point. What is important is for people to understand the probabilistic nature of these tests and of all medical testing. The relative risk for coronary heart disease may be estimated on the basis of genetic variants, but also on the basis of LDL levels. What is important to understand is that people with certain genetic variants and with higher LDL levels are more likely to develop coronary heart disease. The case is clearer for LDL levels because their impact on coronary heart disease has been studied for decades and, most importantly, because the underlying biological mechanism is now known. But even in this case, the connection is always probabilistic. Some of those with high LDL levels may develop coronary artery disease relatively early in life, whereas others may not. Probabilities only indicate how likely an event is to happen, not whether or not it will happen.

A more recent study compared the methods of 23andMe, deCODEme, and Navigenics through a simulation. The study was conducted in a hypothetical population of 100,000 individuals on the basis of published genotype frequencies. The risks for developing particular diseases were calculated using the companies' methods as described on their websites. These diseases were age-related macular degeneration, atrial fibrillation, celiac disease, Crohn's disease, prostate cancer, and type 2 diabetes, which were chosen because they differ in the effect of SNPs and in the average population risks. Age-related macular degeneration and celiac disease are influenced by a few SNPs with strong effects on disease risk, whereas the other diseases are influenced by many SNPs with relatively weak effects. In addition, celiac disease and Crohn's disease are rare disorders, whereas the others are more common. The results of this study showed a substantial difference in the predicted risks of the three companies because of differences in the sets of the SNPs used and in the average population risks they had chosen (Table 6.1), as well as in the formulas used for the calculation of risks. Given the choices made by the companies, as presented in Table 6.1, the differences in the estimated risks in Table 6.2 should not be surprising. These findings show that the results from DTCG tests should be considered with caution, as for certain diseases the risk predictions are more similar among the three companies (e.g., celiac disease) than in others (e.g., prostate cancer).

A poll among readers of the prestigious *New England Journal of Medicine* produced interesting findings. The poll presented the case of a fictional asymptomatic 45-year-old man who was concerned about his risk for cancer. The views of two experts were presented on whether or not they would recommend genetic screening in an asymptomatic person, and in case they did whether the whole genome or just a set of cancer-related genes should be sequenced. On the basis of this, readers of the journal were asked for their recommendations. Of the 929 readers who voted, 47 percent would recommend sequencing for cancer genes only and 12 percent would recommend sequencing the whole genome. The remaining 40 percent would not recommend any genetic testing. Nevertheless, from those readers who would recommend testing and provided comments, most expressed reservations. In particular, they questioned the appropriateness of genetic screening of any kind and suggested that additional information such as family history would be necessary for a better assessment. These results certainly reflect some concerns about genetic testing and about genome sequencing in asymptomatic persons who do not have a strong family history of diseases for which there are known genetic risk variants and available treatments.

Results like these should make consumers concerned about the conclusions that can actually be drawn from these tests. They also set the demand for

Diseases	Average population risk (%)			Number of SNPs used		
	23andMe	deCODEme	Navigenics	23andMe	deCODEme	Navigenics
Macular degeneration	6.5	8	3.1	3	6	6
Atrial fibrillation	27.2	25	26	2	6	2
Celiac disease	0.12	1	0.06	4	8	10
Crohn's disease	0.53	0.5	0.58	12	30	27
Prostate cancer	17.8	16	17	12	26	9
Type 2 diabetes	25.7	25	25	11	21	18

Source: Data based on Kalf et al.'s "Variations in predicted risks in personal genome testing for common complex diseases," p. 87

Table 6.1 Average population risks and number of SNPs used by 23andMe, deCODEme, and Navigenics in the prediction of risks for six multifactorial diseases

Diseases	Assigned to the same risk category by all three companies			Assigned to the same risk category by two companies			Assigned to different risk categories
	↓↓↓	−	↑↑↑	↑↓− / ↓↓−	↑ / ↓−	↑↑↓ / ↓↓↑	↑−↓
Macular degeneration	52.3	0.5	15.2	6.1	6.0	12.5	7.4
Atrial fibrillation	42.4	6.7	16.7	27.3	5.7	1.2	0.0
Celiac disease	75.3	0.0	13.8	9.0	0.4	1.3	0.3
Crohn's disease	51.8	0.2	3.5	13.8	3.7	19.9	7.2
Prostate cancer	15.6	4.5	13.5	29.4	21.7	6.5	9.0
Type 2 diabetes	22.2	7.8	14.7	24.1	23.1	3.2	5.0

Note: According to the risk categories used by 23andMe, which categorizes disease risks as decreased (↓), increased (↑), and average (−) risks if the risks of disease are lower than 20 percent below the average population risk, higher than 20 percent above the average population risk, and in between, respectively (values are percentages). For example, ↓↓↓ indicates the percentage of individuals who were at decreased risk according to all three companies, and ↑−↓ indicates the percentage of individuals for which the three companies predicted risks in all three different risk categories.

Table 6.2 Agreement among the three companies in assigning individuals to the same risk category

regulations that would protect them, and also ensure that they understand the actual conclusions that can be drawn from them. In August 2013, 23andMe launched a national television commercial that suggested that people could learn a lot about their health and risk for disease for $99 (www.ispot.tv/ad/7qoF/23-and-me). But in November of the same year, the Food and Drug Administration (FDA) ordered 23andMe to stop marketing the respective kit until it received authorization, and the company complied. This happened because 23andMe failed to respond to questions from the FDA about the analytical validity (whether the test could accurately detect whether a genetic variant is present or absent) and the clinical validity (how well a genetic variant is related to the presence, absence, or risk of a specific disease) of that test. The FDA requires that companies selling health-related medical devices to the public demonstrate that these are safe and effective. Whereas collecting saliva and sending the sample for DNA analysis does no harm to the individual, the interpretation of the results and the handling of the respective data might do. In response to the FDA's order, 23andMe started providing not health-related genetic risk assessments, but only genetic data (i.e., genotypes) and ancestry-related genetic reports. However, on March 6, 2018, the FDA gave authorization to 23andMe to market a DTCG test for three *BRCA1* or *BRCA2* mutations.

It is also important to remember that it is possible that a genetic test indicates that one has a disease-related variant, whereas this is not the case – this is a "false positive" result. A genetic test may also fail to identify a disease-related variant that one indeed has – this is a "false negative" result. These cases are of course rare, or at least they are less probable than an accurate test. Nevertheless, they exist. The probabilities of a "false positive" or a "false negative" result are reflected by two important characteristics of the test performed: its *sensitivity* and its *specificity*. The sensitivity of a test indicates the probability for a test to yield a true positive result. For example, if the sensitivity of a test is 99 percent, this means that 99 out of 100 tests performed will accurately indicate that these people have the respective variant; however, one of them will not, and this result will be a false positive. The specificity of a test is the probability that a negative test result is a true negative one. For example, if the specificity of a test is 99 percent, this means that 99 out of 100 tests performed will accurately indicate that these people do not have the respective variant; however, one of them will have it, and this result will be a false negative (two related concepts are: the *positive predictive value*, which is the proportion of people who carry the variant among those with a positive test result; and the *negative predictive value*, which is the proportion of people who do not carry the variant among those with a negative test result).

There are additional issues with the validity of DTCG tests. Researchers looked at raw DNA data provided by companies performing these tests, coming from people from whom the companies requested further testing. The researchers found that 40 percent of DNA variants in several genes reported in the raw data were false positives. Furthermore, eight variants in five genes were designated as "increased risk" in the companies' raw data; however, several clinical laboratories had classified these variants as benign, and publicly available databases mentioned them as common variants. The researchers noted that self-tests are less reliable than they should be, and their findings should be confirmed in clinical laboratory settings that have the required expertise in the detection and classification of complex variants. The researchers noted that "While the raw data include disclaimers stating that they have not been validated for accuracy and are therefore not intended for medical use, they could easily be misinterpreted or misused by a consumer or medical provider with little to no training on the complexities of genetics."

The general conclusion that most variants make minor contributions to the development of characteristics or disease has made some researchers think that it might be possible to estimate the cumulative contribution of many variants and therefore estimate the total risk for developing a disease. This cumulative contribution can be estimated with the polygenic risk scores (PRSs). This is the weighted sum of the number of risk alleles carried by an individual, where the risk alleles and their weights are defined by the loci and their measured effects as detected by GWAS. This simply means that if we know that a certain individual carries DNA variants *A, B, C, D*, etc., and if we know from GWAS what the relative risk (probability) is to develop the characteristic of disease for each of these alleles, we can estimate the cumulative risk as the aggregated contribution of all these DNA variants. Researchers select DNA variants that have been found to be associated with a condition in multiple GWAS, and develop a risk model that predicts the cumulative effect of many gene variants. Based on such models, and on an individual's specific genotype, it is possible to calculate an individual's risk for the condition.

This might sound simple and straightforward; however, there are several limitations that one should keep in mind. First, there are no universally agreed upon standards for PRS models, and so it is possible that there exist different approaches for predicting risk for the same condition. As a result, one model may be based on fewer than 100 DNA variants, whereas another may be based on thousands. Another issue is that the models should reflect the disease process: some diseases emerge early in life, whereas others develop later on. Therefore, PRS models should take into account that in the later cases, disease risk increases with age and is not constant throughout life. A third issue is whether PRSs accurately reflect the risk of developing the condition in the general population. Some GWAS tend to distinguish simply between individuals exhibiting the condition (cases) and individuals not exhibiting the condition (controls), whereas it is possible for some people in the latter group to actually develop the condition at some later time. The point here is that many conditions exist not as dichotomous disease vs. no-disease states, but on a continuous spectrum of disease (for instance from glucose intolerance to prediabetes to type 2 diabetes). Therefore, PRSs can be useful only if they indicate an individual's absolute risk and not the relative risk compared to a presumed control group. A final issue is that most GWAS have been conducted in populations of European descent and so generalizations of their findings are not possible.

Finally, a very important issue is how people react to and understand the results of genetic tests. In one study, researchers used data from GeneScreen, an exploratory project that targets 17 genes related to 11 medically actionable conditions, to investigate how people perceive and understand genetic tests. Among 1,086 participants recruited, 263 enrolled in the study. Among them, 14 were found to have a positive result (5 of whom were aware of their genetic condition), whereas the other 249 received negative results. The researchers found that 72.5 percent of GeneScreen participants reported personal or family histories of eight health conditions tested. This means that many participants decided to get involved in the study because of actual or perceived risk for one of the conditions tested, and therefore that the sample tested is not representative of the total population. Furthermore, whereas more than 85 percent scored themselves very high for all three questions assessing perceived understanding, less than 20 percent of them demonstrated a good understanding of the standard limitations of a negative test. Finally, participants largely stated that they did not regret taking the test and that they would not change their health-related behaviors in response to their results.

What is the conclusion from all these discussions about genes and personalized medicine? All we can do at this point is compare one's estimated risk with an average population risk for a certain disease. On the one hand, being above the average population risk does not entail that one will necessarily develop a disease; therefore, one should not feel destined to have the disease despite currently being healthy. On the other hand, being below the average population risk does not entail that one will not develop the disease; therefore, one should not feel free to smoke or be exposed to sunlight just because one's estimated genetic risk for lung or skin cancer is low. Most importantly, even if a person carries a DNA variant and that does cause a disease, we have to ask whether there is any medical action that we can take (this is called *clinical utility*). In most cases, there is not (*BRCA* genes might be an exception, as the Angelina Jolie case discussed in Chapter 1 shows), and therefore genetic testing for these seems unnecessary. This of course will likely change in the future, but for now regular medical check-ups are more useful than estimating genetic risks. Therefore, it might be better to worry about what we can currently do, such as adopt a healthy diet, do physical exercise, quit smoking, drink sensibly, avoid prolonged exposure to sunlight, and stop worrying about the rest, even if they are inside us.

To put it simply: Finding out which alleles of particular associated genes we do or do not carry is not really useful, with the exception of a few genes, as at this point there are not particular actions we can take or medical interventions we can undergo.

Genes Are Not "Texts" Waiting to be Read: Beware of Gene Metaphors

Metaphors are very common in science because in order to represent or explain natural entities, phenomena, processes, or mechanisms that we do not fully understand, it is often easier to refer to something else that is familiar and considered to have similar characteristics. Simply put, a metaphor is understanding and experiencing something in terms of something else with which we are already familiar. It can be thought of as a mapping from a source domain of everyday experience to a target domain, with the aim of better understanding the latter in terms of the former. This is done by emphasizing particular features of the source domain and hiding others. As evolutionary geneticist John Avise has nicely put it: "Evocative metaphors can distill an ocean of information, whet the imagination, and suggest promising channels for navigating uncharted ... waters."

Metaphors have indeed been useful in genetics and genomics research. This has certainly been the case for the metaphor of "information" encoded in DNA that paved the way for deciphering the genetic code, the metaphor of the genome as the "book of life" on which the Human Genome Project (HGP) was based, the metaphor of the "encyclopedia" employed by the researchers of the ENCODE project, or the program metaphor used by Craig Venter to describe life as "a DNA software system" that both creates and directs the (more visible) hardware of life, such as proteins and cells. Several other metaphors have been employed: DNA analysis has been described as "reading"; DNA replication has been described as "copying"; RNA synthesis has been described as "transcription"; protein synthesis has been described as "translation"; RNA modification has been described as "editing"; and more. Such metaphors are not inherently wrong, and can actually help us make sense of the respective phenomena. But nonexperts especially should always keep in mind that metaphors are a means of representation, nothing more. "Books," "software," "reading," "writing," and so on are all human inventions and thus have an inherent dimension of anthropomorphism. This needs to be made explicit, or we should otherwise avoid any unnecessary use of expressions of this kind.

However, two kinds of problems can emerge. The first is that it is often very easy to confuse the target domain with the source domain. That is, it is often very easy to forget that the properties and the features of the source domain attributed to the target domain are not really its own properties and features. For instance, in the example of "signaling" and "receptor" proteins, one should keep in mind that signaling proteins are not really active agents in intercellular communication, do not have any intentions, and their interactions and subsequent changes depend on several other molecules around them. Most importantly, there is really no signal transfer as it would happen in an electrical circuit. The second problem is that the focus may eventually come to be only on those features of the target domain in which the metaphor is better illustrated, and thus overlook other, perhaps important, aspects that do not fit well in the metaphor. In the case of the signaling–receptor interaction, the problem with focusing on the interaction between the two proteins too much is that it may mask the importance of the broader cellular context in which the chemical interactions between these two proteins take place. This is why it is necessary to be aware both of the specific features of metaphors and of their different uses. If this is not done, then metaphors can cause significant problems both within science and in science communication.

Genes can be considered as an exemplar case of the negative impact that the bad use of metaphors and the use of bad metaphors can have. Perhaps most influential has been the metaphor of "gene action," introduced in the beginning of the twentieth century. This metaphor has been unquestionably productive, but also problematic because it facilitated the attribution of agency, autonomy, and causal primacy to genes. By not knowing what a gene is and by talking at the same time about gene action, it became possible to consider genes as the basis of life. Genes have thus often been described as autonomous agents producing phenotypes, which is another anthropomorphic metaphor that may imply that genes do things on their own. But this is, of course, inaccurate. It is therefore important to reflect upon the limits of the metaphors we use. We can say that genes "encode" some "functional" products, insofar as we clearly explain that this is just a way of representing the informational properties of DNA. These are not inherent, and make sense only in the cellular context in which they can in turn be used as a resource for the production of molecules that contribute to the maintenance and the roles of self-regulated, living systems. For this reason, it might be better to replace the concept of gene action with that of gene interaction. This means we should refrain from talking about genes that do this or that, and refer to genes that interact with other genes and with their environment. With simple changes like these we can at least give a better sense of the complexity of these phenomena.

A useful metaphor about genes was provided by molecular biologist Robert Pollack. He suggested that we had better think of the human genome in terms of particular linguistic properties, syntax, grammar, and semantics. In this view, DNA nucleotides are letters that combine to form words, which in turn combine to form sentences: the various alleles of the same gene. These sentences do not form a book that should be read from beginning to end; rather, they form a lexicon, a collection of arbitrary ordered sentences on which the cell draws for information – in the same sense that we use an encyclopedia with an arbitrary alphabetical order of entries. To understand these entries, Pollack argued, we need to be able to understand the words because "no linear sequence, whether of DNA, RNA, or amino acids, can reveal the meaning of a protein any more than we might grasp from its letters the meaning of a new word in a foreign language." But this requires more than simply reading the DNA sequence. It requires making sense of the respective cellular processes as a whole.

Overall, useful as they might be, metaphors have limitations. Often metaphors are used because we ignore the details and so they have a heuristic value both in explaining the respective phenomena and in guiding further research. It seems that we will have to rely on metaphors, and that for various reasons they cannot be avoided. Therefore, what we need to do is keep their limitations in mind, and scientists should be explicit about those.

Let us consider one metaphor in detail: "gene editing." Already in the early 1960s researchers had concluded that small, circular DNA molecules, called plasmids, could be transferred among bacteria, conveying – among other properties – resistance to antibiotics. This was a major issue, as drug-resistant bacteria could be a cause of epidemics. Therefore, scientists needed to find ways to replicate plasmids and cut them apart in order to understand how they acquired and transferred resistance to antibiotics. It was soon found that genes conferring resistance to antibiotics were naturally recombined in the plasmids. They also figured out how to make bacterial cells take up plasmids, as well as how to use restriction enzymes to cut plasmids at precise sites and insert other DNA segments therein using DNA ligases. They could thus insert particular genes to particular plasmids, producing recombinant plasmid molecules, and then select only those bacteria that had taken up the recombinant DNA. This is how the era of recombinant DNA began, through the work of scientists such as Paul Berg, Naomi Datta, Ephraim Anderson, Herbert Boyer, and Stanley Cohen. Once the technology worked, it raised both hope and concerns. The scientists themselves became concerned about the applications of recombinant DNA, organizing most famously the Asilomar conference in 1975. The recombinant DNA technology was expanded over the following decades, but its main principles remained the same.

If we think of genes as sentences that consist of words that in turn consist of nucleotides, then we can think of the recombinant DNA processes as the "copy" or "cut" and "paste" procedures we can perform with a word processor on our computer. What is more important, is that scientists cannot only "copy" and "paste" a DNA sequence within the same book; they can also do this from one book to another, even a very different one. In the early 1980s, it became possible to insert the DNA including the human insulin gene into plasmids, which were then taken up by bacteria which eventually expressed the human gene and produced human insulin. That was the first product of the recombinant DNA technology by the company Genentech. In other words, not only text from the human lexicon was copied and pasted into the bacterial lexicon, but it was also possible for bacteria to draw on this information and produce the same molecule that human pancreatic cells also produce. Of course, this does not mean that all such attempts worked as well. The DNA of bacteria is small and they can take up recombinant plasmids; but doing the same procedure with other organisms, such as mammals, is a lot more difficult. Transgenic organisms – those that contain artificially introduced DNA from a different organism – do not necessarily exhibit the properties for which they have been genetically modified.

An important problem of these transgenic DNA technologies is that it is difficult to control where exactly the copied text will be pasted. In the small genome of bacteria this might be less problematic, but in the mammalian genome it could be disastrous. It is one thing to paste a text where you want and another to paste it somewhere, where it disrupts a message that was already there. In the early 1980s, it also became possible to make similar modifications in mice. By injecting DNA into fertilized eggs that were then implanted into female mice, it was possible to produce phenotypic changes in the offspring. It was also found that the mammalian cells themselves could integrate recombinant DNA through the process of crossing over: The "foreign" DNA would pair with a similar DNA already in the genome, and it was possible for DNA segments to be exchanged between the two. This is described as homologous recombination. In this way, a "wrong" DNA sequence in the cell could be replaced by a "correct" one prepared by humans. Even though this had potential not only for research but also for therapeutic purposes, it did not work well. The reason was that nonhomologous recombination was also possible. The "foreign" DNA might pair with another DNA sequence, not the one it was intended to pair with, and eventually be integrated in another place, potentially causing problems. These methods were generally described as "genetic engineering." They lacked precision and did not really allow for making minor changes in single words or even single letters. In the early 1990s, this kind of technology was used for gene therapy in humans. However, success was limited.

All this changed in 2012, when Emmanuelle Charpentier, Jennifer Doudna, and their colleagues published a paper presenting a method for gene editing. They were studying a system naturally occurring in bacteria, called CRISPR-Cas9, which stands for clustered regularly interspaced short palindromic repeats and CRISPR-associated (Cas) protein 9 (this is described as a "scissors" enzyme because it cuts the DNA). The CRISPR are sequences found in bacteria that exhibit a specific pattern: the same short DNA sequence (hence short) was repeated several times (hence repeats) one after the other (hence clustered); these sequences were the same when they were read from left to right or right to left (e.g., GATTAG, hence palindromic); and each of these palindromic sequences was separated from the ones before and after them by other short DNA sequences (hence regularly interspaced) which were different from one another, which we can call spacers. It was later found that these spacers came from viruses that had infected bacteria. Those bacteria that survived had copied parts of the viral genome and had inserted them in their own genome in the form of these spacers. If a virus infected anew the bacterium, these spacers were copied and the copies targeted the viral genome and facilitated its destruction by enzymes.

Doudna and Charpentier realized that this system could be used in genetic engineering procedures, and so developed the method CRISPR-Cas9, which is generally known as "gene editing." Here is how this works: A single stranded "guide" RNA molecule is chemically synthesized, consisting of two specific RNA sequences. One of them is CRISPR RNA; this is complementary to the sequence of the gene that researchers want to modify. The "guide" RNA molecule can thus bind to one of the strands of the DNA molecule of the cell when the two DNA strands are separated because of its complementary sequence. The other specific RNA sequence in this "guide" RNA molecule is tracrRNA; this can be recognized by the "scissors" enzyme that is capable of cutting the DNA close to the DNA sequence to which the "guide" RNA matches. The repair mechanisms of the cell rejoin the cut ends of the DNA molecule, thus leaving out the DNA segment that matched the "guide." In this way, the DNA sequence can be changed. Even though scientists cannot control how exactly the DNA is repaired by the cell mechanisms, the procedure works quite well. The most basic application of this procedure is to disrupt the sequence of a gene so that the corresponding protein is not produced or is altered and so cannot do what it normally does. Homologous recombination can also be used for replacing a harmful DNA sequence with one that does not relate to that disease. In these cases, a precise "repair" is possible.

Here is how Jennifer Doudna described the method in her popular science book on the topic: "the genome – an organism's entire DNA content, including all its genes – has become almost as editable as a simple piece of text"; "Because CRISPR allows precise and relatively straightforward DNA editing, it has transformed every genetic disease – at least, every disease for which we know the underlying mutation(s) – into a potentially treatable target." There are two important metaphors here: "editing" and "targeting." The big question is: Are these metaphors accurate? Or are they misleading because they represent the respective phenomena as less complex than they actually are, and because they exaggerate the control that scientists have over these procedures? To provide answers to these questions, it is necessary to understand where these metaphors come from.

According to language and metaphor scholar Brigitte Nerlich, "gene editing" is rooted in an old metaphor of the genome as the "book of life." Therefore, it is important to consider the issues related to this metaphor. As we saw earlier, this is a metaphor that should be used with caution because in the best case this book is a lexicon with a collection of arbitrary ordered sentences. As Nerlich noted:

> As we have seen, the hopes, hypes and concerns surrounding the book of life metaphor have remained almost constant over time, while, at the same time, the metaphor has moved closer to reality. However, we should still be careful to not confuse hype with reality. The book of life will always be complex, complicated and messy, and reading, writing or editing it will never be as straightforward as it might appear to be or to become. Metaphors like "the book of life" or "genome editing" are useful in encapsulating all this complexity, but they can only afford us glimpses of what's going on. They should not be taken as literal representations.

This is a very important point. It is one thing to use the book metaphor to produce a representation of the genome, and another to come to believe that reading or editing the "book of life" is as simple and straightforward as reading an actual book.

But are these metaphors used in the public sphere? A study looked at the use of metaphors associated with CRISPR in some widely read US newspapers and popular science publications between January 2013 and July 2015. Using as their search criterion any appearance of the term "CRISPR," the researchers found 22 newspaper articles and 24 popular science articles. The metaphor of "editing" was found in 21 of the 22 articles in the newspapers, and in 20 of 24 articles in popular science publications, appearing 168 times overall. Importantly, these editing metaphors presented genomes as text to be edited, without conveying any sense of risk when discussion was about genomes in general. But the situation was different when discussion was about editing human embryos, where concerns were raised. The metaphor of "targeting" appeared 56 times overall, in 12 of 22 newspaper articles and in 12 of 24 popular science articles. In this case, discussion was both about precision and about the danger of unintended outcomes, thus raising concerns about the dangers of this technology. In other words, the uses of the "targeting" metaphor were clearer about the potential dangers than those of the "editing" metaphor.

Are the dangers real? As I am writing these lines in September 2020, a report by the International Commission on the Clinical Use of Human Germline Genome Editing, the National Academy of Medicine, the National Academy of Sciences, and The Royal Society was published, with the title *Heritable Human Genome Editing*. The motivation for producing this report was the announcement in 2018 of the use of heritable human genome editing (HHGE) by He Jiankui in China, resulting in the birth of children whose DNA had been edited. The first recommendation of this report requires no further comment:

> Recommendation 1: No attempt to establish a pregnancy with a human embryo that has undergone genome editing should proceed unless and until it has been clearly established that it is possible to efficiently and reliably make precise genomic changes without undesired changes in human embryos. These criteria have not yet been met and further research and review would be necessary to meet them.

In late June 2020, three (at the time not yet peer-reviewed) articles appeared that reported problems with the CRISPR method when used to modify human embryos. In all studies, the embryos were used for scientific purposes only, and not to generate pregnancies. The studies showed that this method can bring about large, unwanted changes to the genome, such as large DNA deletions and reshuffling, at or near the target site. In the first study, about one in five embryos that had undergone genome editing had unintended DNA rearrangements and large deletions. In the second study, about half of the embryos tested had lost large segments of the chromosome, and sometimes the whole chromosome, on which the targeted genes were located. Finally, the researchers conducting the third study also found that editing had affected large regions of the chromosome on which the targeted gene was found. Shouldn't "editing" and "targeting" metaphors be clear about these kinds of dangers too?

To put it simply: We should beware of gene metaphors as they may attribute to genes more power than they actually "have." Worse, metaphors such as "gene editing" may make us think we can know and do more than we actually can. The HGP, the ENCODE project, and GWAS have unquestionably confirmed this point: One should look at the genome as a whole, and not at individual genes, to understand phenomena. Gene metaphors are misleading if they do not convey this message.

Concluding Remarks: How to Think and Talk about Genes?

In the preface of the present book I mentioned the 1997 film *GATTACA*. What was science fiction then is nowadays represented as a possibility. Private companies seem to offer *GATTACA*-type genetic tests to parents, with the promise to assess embryos for a variety of conditions. Despite the limitations of polygenic risk scores (PRSs), briefly discussed in the previous chapter, according to the company Genomic Prediction, they can be used to assess disease risk. To achieve this, couples have to undergo the procedures of *in vitro* fertilization. Each woman receives hormones to stimulate ovulation, and ova are selected and fertilized with their husband's sperm. Then, after a few days, embryos consisting of a handful of cells undergo preimplantation genetic diagnosis. This is a process during which a single cell of the developing embryo is removed and tested for the presence of DNA variants related to a condition. On the basis of this, PRSs for each embryo can be calculated and decisions can be made about which embryo(s) should be transferred to the mother's uterus for implantation. Currently, the company is quite clear on their website about what can and cannot be done.

According to Stephen Hsu, cofounder of Genomic Prediction, the company is currently focused on "serious" diseases. "Anything that we are putting warnings in the report about are things that are medically classified as diseases. We are just informing the physician or the genetic counselor that this embryo is at elevated risk for a particular disease or normal risk. But we don't go beyond that." This implies that the company does not aim at providing advice to couples and their genetic counselors; they only provide their risk estimation, and they only do this for disease. Hsu added that

> Away from diseases, we can predict cosmetic traits really well, so we can figure out who is blonde, redhead, blue eyes; who has light skin, dark skin … But even though there is high demand for it, we don't plan to do cosmetic stuff at Genomic Prediction. The non-disease traits, like height, I think we only warn parents if the child is in danger of idiopathic short stature, which is a medical condition.

The related press coverage in autumn 2019 attracted criticism from many experts at that time. Studies have also shown that screening human embryos for polygenic characteristics has limited utility. For instance, one study that used simulated data concluded that the top-scoring embryos would be expected to be about 2.5 cm or 2.5 IQ points above the average, whereas some children that would have had high scores would not be the tallest in their families. Perhaps the hope that we can "read" our genome, "the book of life" and make predictions about future outcomes is something people tend to find intuitive and accept unquestionably. But this is far from possible, at least for now.

A main aim of the present book has been to show that there are no "genes for" characteristics or disease. But the idea of a single "gene for" a characteristic or disease is untenable even at a theoretical level. The attribution of characteristics and diseases to DNA makes no sense conceptually because it is not DNA that is directly responsible for them. It is proteins of different kinds that directly affect characteristics or diseases. Therefore, it would make more sense to state that a person has a certain (defective) hemoglobin "for" β-thalassemia. But even such a statement would again be inaccurate because several enzymes and other proteins are involved in the synthesis of any given protein in our cells. As each of these enzymes and proteins is synthesized on the basis of some genes, all those genes are relevant to the synthesis of, for example, β-globin and not just the *HBB* gene that encodes it. Interestingly, the term "protein" is derived from the Greek word "πρωτεῖος" ("πρῶτος" in modern Greek), which means first in rank or position. It seems that we had to unravel the secrets of DNA in order to eventually realize, and perhaps reaffirm, the importance of proteins.

But why such an interest in genes? Richard Lewontin has argued that we have confused "the methodological limitations of experiments" with "the correct explanations of phenomena." Because of how easily we can experimentally induce major genetic changes that have large effects, we have come to equally easily accept the belief that genes determine organisms. Developmental biologists, Lewontin argued, ask questions about the differentiation of the anterior and posterior parts of animals, and the formation of the body parts in between those, because they have found that single gene changes can bring about significant alterations in these processes. But they do not know how to ask questions about why different individuals have heads and legs of different sizes and shapes, and so they never ask these questions. As the old joke goes, it is as if one is looking for one's keys under the light where one can see better, independently of where one lost them. During the twentieth century, we managed to control genes and so we initially thought that they are the absolute determinants of organisms. However, research in the twenty-first century has revealed the complexities of organismal development, and we have now realized that we also have to look for the keys in the dark because we could not find them under the light.

In Chapter 1, I mentioned the term "geneticization," which was coined in 1991 by epidemiologist Abby Lippman to describe the phenomenon of making overt attributions to genes. Thirty years later, have we seen the geneticization of society? A recent analysis of the literature on geneticization since the early 1990s has concluded that little of what Lippman had anticipated actually materialized. As I also showed in the present book, the contributions of genetics and genomics to clinical practice have been limited so far. However, at the same time it seems that a powerful genetic imaginary persists, which continues to fuel expectations for future advancements. This is evident in the number of people who undergo genetic testing despite the uncertainties and the limitations. Why? Because even though there was no geneticization of society, genetic fatalism is an intuitive view: It makes sense to believe that genes are our essences, which determine and best explain who we are. People still look to genes, however perceived, and DNA for certainty. But DNA data can be used to estimate probabilistic outcomes and not to predict, as well as to point to a certain conclusion but not without possible pitfalls. There are two main reasons for this.

The first are the current practical limitations. This may be overcome in the future, of course. In fact, if one considers the research tools that Morgan and his collaborators had 100 years ago, today's methods and techniques would seem unimaginable back then. Therefore, whatever seems to be science fiction today could become everyday practice in the future. But we are not there yet. Therefore, we need to be aware of current abilities and make sure not to distort them. This means that we should neither exaggerate what can be done, nor downplay it. Until the moment that these lines are written, we can sequence genomes and we can find associations between DNA variants and characteristics or disease. Nothing more. The validity of the respective methods and approaches are continuously increasing, but their clinical utility is still limited. Twenty years after completing the "script" of the human genome we are still struggling to decode it. We have learned to read it, but not yet to fully interpret it. But it seems that we are on the right track.

The second reason is a limitation inherent in human nature. We look for certainty in a probabilistic world, where combinations of critical, unpredictable events shape outcomes. We will never find it! We need to realize and accept that data become evidence only in the light of human inferential abilities, and so any conclusion depends on them. DNA tests tell us nothing about the prospect of each one of us developing a disease; they only make probabilistic estimates. Our DNA analyses – which are not infallible – can produce data from which we can draw inferences – which, again, are not infallible. Our efficiency in doing DNA analyses and in drawing inferences from DNA data are continuously improving. But there will always be limitations. This is why researchers and scientists who communicate research and its findings to society should be very careful about what language they use. This language should reflect both the actual potential and the actual limitations.

Why do we consider genes to be so important? For some reason, we seem to think that there is something inside us that determines our essence: who we are. Apparently, if organisms have essences they must be at some deep level – this is by definition what an essence is about. But as I have explained so far, genes cannot be these essences. In contrast, we can think of the developmental capacities of organisms as their essences. The capacity to develop on the basis of a particular genetic material expressed under particular environmental conditions, exhibiting both robustness and plasticity, could be perceived as the essence of organisms. It is therefore time to reconsider our conception of how we come to be as we are, and to think of developmental processes as important factors in this process instead of genes alone. As explained in this book, genes are implicated in development and account for differences in characteristics, but depend on their cellular context and a lot more that exists within our genome. Therefore, it is important to understand the complexity of all these phenomena in order to realize that genes are not our essences, that they do not determine who we are, and that they cannot alone explain characteristics and diseases.

So, how should we think about genes? Here is the take-home message of this book: Genes were initially conceived as immaterial factors with heuristic value for research, but along the way they acquired a parallel material identity as DNA segments. The two identities never converged completely, and therefore the best we can do so far is to think of genes as DNA segments that encode functional products. There are neither "genes for" characteristics nor "genes for" diseases. Genes do nothing on their own, but are important resources for our self-regulated organism. If we insist on asking what genes do, we can accept that they are implicated in the development of characteristics and disease, and that they account for variation in characteristics in particular populations. Beyond that, we should remember that genes are part of an interactive genome that we have just begun to understand, the study of which currently has various limitations. Genes are not our essences, they do not determine who we are, and they are not the explanation of who we are and what we do. Therefore, genetic fatalism is simply wrong: We are not prisoners of any genetic fate, with the exception of people who suffer from particular cases of genetic disease. This is what the present book has aimed to explain.

If you agree with this and if you are now able to explain this to others and help them overcome the intuitive conception of "genes for," I think you should be confident that you understand genes.

Summary of Common Misunderstandings

Common misunderstandings about genes and responses stemming from this book:

Genes can be structurally individuated on the DNA of chromosomes. Genes are not distinct segments of DNA. They may overlap and each one of them may produce different functional products, which in turn are those that seem to determine which sequence is a gene.

Genes are fixed entities, which are transferred unchanged across generations and which are the essence of what we are by specifying characteristics from which their existence can be inferred (genetic essentialism). Genes are not fixed and may change as they are transferred across generations due to mutations or recombination events. Most importantly, they are not the essences of what we are as they cannot alone specify our characteristics. If we have essences, these are developmental ones, depend on genes, environmental factors, and their interactions, and they are characterized by both robustness and plasticity.

Genes invariably determine characteristics, so that the outcomes are just a little, or not at all, affected by changes in the environment or by the different environments in which individuals live (genetic determinism). Genes do not determine characteristics in any strict sense. Rather, genes encode information for the development of characteristics, which cells use as a resource during their multiplication and differentiation in the course of developmental processes and always in combination with environmental signals. In fact, the contribution of both genes and environment is necessary for development and they are so intertwined that it is impossible to distinguish between the two at the individual level.

Genes provide the ultimate explanation for characteristics, and so the best approach to explain these is by studying phenomena at the level of genes (genetic reductionism). Genes cannot provide the ultimate explanation for characteristics, as the latter are the outcomes of complex developmental processes, not of genes alone. Of course, changes in genes can bring about (and therefore explain) changes in characteristics.

There exist "genes for," that is, genes that cause particular characteristics or diseases. There are no "genes for" characteristics or diseases simply because their relation is a many-to-many one. Many genes are implicated in the development of a characteristic or a disease, and each gene can be implicated in the development of several characteristics or diseases. Even in those cases where there is a strong connection between a gene and a characteristic or a disease, there are always several others that are directly or indirectly involved.

Genomes are the sum of genes. Genomes consist of several other sequences besides genes, which are involved in the expression and regulation of genes. Furthermore, these phenomena can be influenced by reversible changes in DNA itself and the histones that surround it, which can occur during the course of one's life, as research in epigenetics has shown.

When a gene is strongly associated with a characteristic or disease, then this gene should be tested for and considered. Association does not equal causation. Even if a gene is clearly associated with a characteristic or disease at the population level, this does not entail that this gene is the only one responsible for phenotypic outcomes at the individual level.

Genomics and sequence technologies have allowed us to read the information encoded in our DNA. Genomics and sequence technologies have allowed us to read the sequence of our DNA, but we are still far from decoding it. Furthermore, genes are not really "texts" waiting to be read; this is a metaphor that we have used to represent our aim to understand genes, but it often masks the fact that it is far from simple and straightforward to make sense of genes.

References

All references appear here in the order that the respective topics are discussed in the main text of the book.

Preface

Scene from movie *GATTACA*: www.youtube.com/watch?v=lP1cCjBkWZU (accessed November 29, 2019).

Estimates for how many people have taken a genetic test: www.technologyreview .com/s/612880/more-than-26-million-people-have-taken-an-at-home-ancestry-test (accessed July 7, 2020); https://blogs.ancestry.com/ancestry/2020/02/05/our-path-forward (accessed July 7, 2020).

Chapter 1

On the definition of scientific expertise: Nichols, T. M. (2017). *The Death of Expertise: The Campaign against Established Knowledge and Why It Matters*. New York: Oxford University Press, pp. 28–39.

On "imagenation": Van Dijck, J. (1998). *Imagenation: Popular Images of Genetics*. London: Macmillan, pp. 11, 198.

Quotation of O'Riordan, K. (2010). *The Genome Incorporated: Constructing Biodigital Identity*. New York: Routledge, p. 14.

Links to the media articles (all accessed on March 18, 2020): https://edition.cnn .com/2013/08/06/showbiz/gallery/hollywood-famous-parents-kids/index.html;

www.ft.com/content/419733b2-e181-11e6-9645-c9357a75844a; www
.sciencedaily.com/releases/2017/01/170123151411.htm; www.dailymail
.co.uk/sciencetech/article-5934673/Being-rich-successful-really-genes-study-
suggests.html; www.thetimes.co.uk/article/scientists-find-24-golden-genes-
that-help-you-get-rich-mfdf8z9jb; www.telegraph.co.uk/science/2019/02/28/
key-happy-marriage-genes-scientists-discover; https://nypost.com/2019/03/07/
this-gene-could-be-the-secret-to-a-happy-marriage-study; www.nytimes.com/
2015/05/24/opinion/sunday/infidelity-lurks-in-your-genes.html?partner=rss
&emc=rss; www.dailymail.co.uk/sciencetech/article-2954349/Women-likely-
cheat-partner-carry-infidelity-gene-scientists-discover.html; www.gene
partner.com

For the definition of geneticization: Lippman, A. (1991). Prenatal genetic testing
and screening: constructing needs and reinforcing inequities. *American
Journal of Law and Medicine*, 17(1–2), 15–50, p. 19.

Quotation of Ruth Hubbard: Hubbard, R., and Wald, E. (1997). *Exploding the
Gene Myth: How Genetic Information Is Produced and Manipulated by
Scientists, Physicians, Employers, Insurance Companies, Educations, and Law
Enforcers*. Boston: Beacon Press, p. 164.

Quotations of Nelkin and Lindee: Nelkin, D., and Lindee, S. M. (2004). *The DNA
Mystique: The Gene as a Cultural Icon*. Ann Arbor: University of Michigan
Press, p. 16, p. 2, p. 204.

Study of magazine articles: Condit, C. M. (1999). *The Meanings of the Gene:
Public Debates about Human Heredity*. Madison: University of Wisconsin
Press.

Study of newspaper articles: Carver, R. B., Rødland, E. A., and Breivik, J. (2013).
Quantitative frame analysis of how the gene concept is presented in tabloid
and elite newspapers. *Science Communication*, 35(4), 449–475.

Study of science fiction films: Kirby, D. A. (2007). The devil in our DNA: a brief
history of eugenics in science fiction films. *Literature and Medicine*, 26(1),
83–108.

Study of science fiction films and novels: Hamner, E. (2017). *Editing the Soul:
Science and Fiction in the Genome Age*. University Park: Pennsylvania State
University Press, pp. 9–11.

Study of television series: Bull, S. (2019). *Television and the Genetic Imaginary*. London: Palgrave Macmillan (quotations from p. 216).

On WEIRD people: Henrich, J., Heine, S. J., and Norenzayan, A. (2010). Most people are not WEIRD. *Nature*, 466(7302), 29.

Global study across 22 countries on public perceptions of genomics: Middleton, A., Milne, R., Almarri, M. A., et al. (2020). Global public perceptions of genomic data sharing: what shapes the willingness to donate DNA and health data? *American Journal of Human Genetics*, https://doi.org/10.1016/j.ajhg.2020.08.023.

On genetic exceptionalism: Middleton, A., Milne, R., Howard, H., et al., on behalf of the Participant Values Work Stream of the Global Alliance for Genomics and Health. (2020). Members of the public in the USA, UK, Canada and Australia expressing genetic exceptionalism say they are more willing to donate genomic data. *European Journal of Human Genetics*, 28, 424–434.

Coriell Personalized Medicine Collaborative: Schmidlen, T. J., Scheinfeldt, L., Zhaoyang, R., et al. (2016). Genetic knowledge among participants in the Coriell Personalized Medicine Collaborative. *Journal of Genetic Counseling*, 25(2), 385–394.

Survey about the genetic and environmental contributions to variation in 21 human traits: Willoughby, E. A., Love, A. C., McGue, M., et al. (2019). Free will, determinism, and intuitive judgments about the heritability of behavior. *Behavior Genetics*, 49(2), 136–153.

Survey with articles on behavior genetics: Morin-Chassé, A. (2014). Public (mis)understanding of news about behavioral genetics research: a survey experiment. *BioScience*, 64(12), 1170–1177.

On people being genetic essentialists: Dar-Nimrod, I., Cheung, B., Ruby, M., and Heine, S. (2014). Can merely learning about obesity genes affect eating behavior? *Appetite*, 81, 269–276; Heine, S. J. (2017). *DNA is Not Destiny: The Remarkable, Completely Misunderstood Relationship between You and Your Genes*. New York: W.W. Norton; Heine, S. J., Cheung, B. Y., and Schmalor, A. (2019). Making sense of genetics: the problem of essentialism. *Hastings Center Report*, 49(3), S19–S26.

Jolie's essays in the *New York Times*: Jolie, A: My medical choice. *New York Times*, May 14, 2013, p. A25; Jolie Pitt, A: Angelina Jolie Pitt: diary of a surgery. *New York Times*, March 24, 2015, p.A23.

Studies on Internet information-seeking queries: Noar, S. M., Althouse, B. M., Ayers, J. W., Francis, D. B., and Ribisl, K. M. (2015). Cancer information seeking in the digital age: effects of Angelina Jolie's prophylactic mastectomy announcement. *Medical Decision Making*, 35(1), 16–21; Juthe, R. H., Zaharchuk, A., and Wang, C. (2015). Celebrity disclosures and information seeking: the case of Angelina Jolie. *Genetics in Medicine*, 17(7), 545–553.

Study of newspaper accounts of Angelina Jolie's double mastectomy: Kamenova, K., Reshef, A., and Caulfield, T. (2014). Angelina Jolie's faulty gene: newspaper coverage of a celebrity's preventive bilateral mastectomy in Canada, the United States, and the United Kingdom. *Genetics in Medicine*, 16 (7), 522–528.

Study of website accounts of Angelina Jolie's double mastectomy: Dean, M. (2016). Celebrity health announcements and online health information seeking: an analysis of Angelina Jolie's preventative health decision. *Health Communication*, 31(6), 752–761.

Studies on the public perception and understanding of Jolie's story: Borzekowski, D. L. G., Guan, Y., Smith, K. C., Erby, L. H., and Roter, D. L. (2013). The Angelina effect: immediate reach, grasp, and impact of going public. *Genetics in Medicine*, 16, 516–521; Abrams, L. R., Koehly, L. M., Hooker, G. W., et al. (2016). Media exposure and genetic literacy skills to evaluate Angelina Jolie's decision for prophylactic mastectomy. *Public Health Genomics*, 19(5), 282–289.

On risk and probabilities: Gigerenzer, G. (2002). *Calculated Risks: How to Know When Numbers Deceive You*. New York: Simon and Schuster.

Data on breast cancer: American Cancer Society (2019). *Breast Cancer Facts & Figures 2019–2020*. Atlanta: American Cancer Society (data and quotations from pp. 4, 13).

Study of women with early-onset breast cancer: Copson, E. R., Maishman, T. C., Tapper, W. J., et al. (2018). Germline BRCA mutation and outcome in young-onset breast cancer (POSH): a prospective cohort study. *The Lancet Oncology*, 19(2), 169–180.

On population screening for *BRCA* genes: King, M.-C., Levy-Lahad, E., and Lahad, A. (2014). Population-based screening for *BRCA1* and *BRCA2*. *Journal of the American Medical Association*, 312, 1091.

On familial breast cancer: Wendt, C., and Margolin, S. (2019). Identifying breast cancer susceptibility genes: a review of the genetic background in familial breast cancer. *Acta Oncologica*, 58(2), 135–146.

Steinberg, D. L. (2015). *Genes and the Bioimaginary: Science, Spectacle, Culture*. Farnham: Ashgate, p. 159.

Chapter 2

On the history of heredity: Müller-Wille, S., and Rheinberger, H-J. (2012). *A Cultural History of Heredity*. Chicago: University of Chicago Press.

On the history of the gene concept: Rheinberger, H. J., and Müller-Wille, S. (2017). *The Gene: From Genetics to Postgenomics*. Chicago: University of Chicago Press; Beurton, P., Falk, R., and Rheinberger, H. J. (Eds) (2000). *The Concept of the Gene in Development and Evolution: Historical and Epistemological Perspectives*. Cambridge: Cambridge University Press; Keller, E. F. (2000). *The Century of the Gene*. Cambridge MA: Harvard University Press.

On the history of Mendel's work: Olby, R. C. (1985). *Origins of Mendelism* (2nd ed.). Chicago: University of Chicago Press.

Mendel's paper: Mendel, G. (2016). Experiments on plant hybrids (1866). Translation and commentary by Staffan Müller-Wille and Kersten Hall. British Society for the History of Science Translation Series, www.bshs.org.uk /bshs-translations/mendel (accessed September 25, 2020).

Johannsen's quotations come from: Roll-Hansen, N. (2014). Commentary: Wilhelm Johannsen and the problem of heredity at the turn of the 19th century. *International Journal of Epidemiology*, 43(4), 1007–1013.

On the work of Morgan and his collaborators: Kohler, R. E. (1994). *Lords of the Fly: Drosophila Genetics and the Experimental Life*. Chicago: University of Chicago Press.

Quotations from Morgan: Morgan, T. H. (1913). Factors and unit characters in Mendelian heredity. *American Naturalist*, 47, 5–16, p. 5; Morgan, T. H.,

Sturtevant, A. H., Muller, H. J., and Bridges, C. B. (1915). *The Mechanism of Mendelian Heredity*. New York: Henry Holt and Company, pp. 208–210; Morgan, T. H. (1917). The theory of the gene. *American Naturalist*, 51, 513–544, pp. 514–515, 517; Morgan, T. H. (1926). *The Theory of the Gene*. New Haven: Yale University Press, pp. 309–310.

On the first evidence about genes being material entities: Muller, H. J. (1927). Artificial transmutation of the gene. *Science*, 46, 84–87, p. 86; Creighton, H. B., and McClintock, B. (1931). A correlation of cytological and genetical crossing-over in *Zea mays*. *Proceedings of the National Academy of Sciences*, 17(8), 492–497.

"Thomas H. Morgan – Nobel Lecture: The Relation of Genetics to Physiology and Medicine," www.nobelprize.org/nobelprizes/medicine/laureates/1933/morgan-lecture.html

Quotation from Beadle and Tatum: Beadle, G. W., and Tatum, E. L. (1941). Genetic control of biochemical reactions in *Neurospora*. *Proceedings of the National Academies of Science USA*, 27, 499–506, p. 506.

On the history of molecular biology: Judson, H. F. (1996). *The Eighth Day of Creation: The Makers of the Revolution in Biology (Commemorative Edition)*. New York: Cold Spring Harbor Laboratory Press; Morange, M. (1998). *A History of Molecular Biology*. Cambridge, MA: Harvard University Press; Olby, R. (1994/1974). *The Path to the Double Helix*. New York: Dover; Kay, L. E. (2000). *Who Wrote the Book of Life? A History of the Genetic Code*. Stanford: Stanford University Press; Cobb, M. (2015). *Life's Greatest Secret: The Story of the Race to Crack the Genetic Code*. London: Profile Books.

On the *Lmbr1/Shh* example: Hill, R. E., and Lettice, L. A. (2013). Alterations to the remote control of *Shh* gene expression cause congenital abnormalities. *Philosophical Transactions of the Royal Society Part B*, 368, 20120357; Griffiths, P., and Stotz, K. (2013). *Genetics and Philosophy: An Introduction*. Cambridge: Cambridge University Press, p. 61.

On the developmental gene concept: Gilbert, S. F. (2000). Genes classical and genes developmental. In Beurton, P. J., Falk, R., and Rheinberger, H. J. (Eds). *The Concept of the Gene in Development and Evolution: Historical and Epistemological Perspectives*. Cambridge: Cambridge University Press, 178–192; Morange, M. (2000). The developmental gene concept. In Beurton,

P. J., Falk, R., & Rheinberger, H. J. (Eds). *The Concept of the Gene in Development and Evolution: Historical and Epistemological Perspectives.* Cambridge: Cambridge University Press, 193–215.

Chapter 3

On alternative splicing: Keren, H., Lev-Maor, G., and Ast, G. (2010). Alternative splicing and evolution: diversification, exon definition and function. *Nature Reviews Genetics*, 11(5), 345–355; Schmucker, D., Clemens, J., Shu, H., et al. (2000). Drosophila Dscam is an axon guidance receptor exhibiting extraordinary molecular diversity. *Cell*, 101, 671–684.

On trans-splicing: Preußer, C., and Bindereif, A. (2013). Exo-endo trans splicing: a new way to link. *Cell Research*, 23(9), 1071.

On overlapping genes: Makalowska, I., Lin, C. F., and Makalowski, W. (2005). Overlapping genes in vertebrate genomes. *Computational Biology and Chemistry*, 29(1), 1–12.

A superb summary of the important findings about the complexities of genes during the 1970s and the 1980s is given in Gros, F. (1991). *Les Secrets du Géne*. Paris: Odile Jacob (quotation from p. 297, emphasis in the original, my translation).

Baltimore on the "brain of a cell": Baltimore, D. (1984). The brain of a cell. *Science 84*, 5(9): 149–151.

Watson on HGP: Watson, J. D. (1992). A personal view of the project. In Kevles, D. J., and Hood, L. (Eds). *The Code of Codes: Scientific and Social Issues in the Human Genome Project*. Cambridge, MA: Harvard University Press, 164–173, pp. 167–168, 173.

Gilbert on HGP: Gilbert, W. (1992). A vision of the grail. In Kevles, D. J., and Hood L. (Eds). *The Code of Codes: Scientific and Social Issues in the Human Genome Project*. Cambridge, MA: Harvard University Press, 83–97, p. 96.

On the history of the Human Genome Project: Gannett, L. (2019). The Human Genome Project. *The Stanford Encyclopedia of Philosophy* (winter 2019 edition), Edward N. Zalta (Ed.), https://plato.stanford.edu/archives/win2019/entries/human-genome; Davies, K. (2001). *Cracking the Genome: Inside the Race to Unlock Human DNA*. New York: Free Press.

Genes do not cause characteristics: Keller, E. F. (1994). Master molecules. In Cranor, C. (Ed.). *Are Genes Us? The Social Consequences of the New Genetics.* New Brunswick: Rutgers University Press, 89–98, p. 90.

On the concept of genetic disease: Keller, E. F. (1992). Nature, nurture, and the human genome project. In Kevles, D. J., and Hood, L. (Eds). *The Code of Codes: Scientific and Social Issues in the Human Genome Project.* Cambridge, MA: Harvard University Press, 281–299.

For an accessible presentation of genome technologies: Snyder, M. (2016). *Genomics and Personalized Medicine: What Everyone Needs to Know.* Oxford: Oxford University Press.

Quotations from Lander: Lander, E. S. (1996). The new genomics: global views of biology. *Science*, 274(5287), 536–539; Lander, E. S. (2011). Initial impact of the sequencing of the human genome. *Nature*, 470(7333), 187–197.

Quotations and information about the ENCODE project: ENCODE Project Consortium. (2004). The ENCODE (ENCyclopedia of DNA elements) project. *Science*, 306(5696), 636–640; ENCODE Project Consortium. (2007). Identification and analysis of functional elements in 1% of the human genome by the ENCODE pilot project. *Nature*, 447(7146), 799.

ENCODE definition of gene: Gerstein, M. B., Bruce, C., Rozowsky, J. S., et al. (2007). What is a gene, post-ENCODE? History and updated definition. *Genome Research*, 17(6), 669–681.

ENCODE on gene function: ENCODE Project Consortium (2012). An integrated encyclopedia of DNA elements in the human genome. *Nature*, 489(7414), 57–74.

On criticisms of the ENCODE project's conclusions about gene function: Eddy, S. R. (2013). The ENCODE project: missteps overshadowing a success. *Current Biology*, 23(7), R259–R261; Graur, D., Zheng, Y., Price, N., et al. (2013). On the immortality of television sets: "function" in the human genome according to the evolution-free gospel of ENCODE. *Genome Biology and Evolution*, 5(3), 578–590; Graur, D., Zheng, Y., and Azevedo, R. B. (2015). An evolutionary classification of genomic function. *Genome Biology and Evolution*, 7(3), 642–645; Doolittle, W. F. (2013). Is junk DNA bunk? A critique of ENCODE. *Proceedings of the National Academy of Sciences*, 110(14), 5294–5300.

Response to criticisms of the ENCODE project: Mattick, J. S., and Dinger, M. E. (2013). The extent of functionality in the human genome. *HUGO Journal*, 7(1), 2.

Latest ENCODE findings: Snyder, M. P., Gingeras, T. R., Moore, J. E., et al. (2020). Perspectives on ENCODE. *Nature*, 583(7818), 693–698.

On GWAS: Tam, V., Patel, N., Turcotte, M., et al. (2019). Benefits and limitations of genome-wide association studies. *Nature Reviews Genetics*, 20(8), 467–484; Kruglyak, L. (2008). The road to genome- wide association studies. *Nature Reviews Genetics*, 9(4), 314–318; Visscher, P. M., Brown, M. A., McCarthy, M. I., and Yang, J. (2012). Five years of GWAS discovery. *American Journal of Human Genetics*, 90(1), 7–24.

On human genetic variation: Frazer, K. A., Murray, S. S., Schork, N. J., and Topol, E. J. (2009). Human genetic variation and its contribution to complex traits. *Nature Reviews Genetics*, 10(4), 241–251.

On multiple causes: Cranor, C. F. (2013). Assessing genes as causes of human disease in a multi-causal world. In Krimsky, S., and Gruber J. (Eds). *Genetic Explanations: Sense and Nonsense*. Cambridge, MA: Harvard University Press, 107–121.

On the concept of postgenomics: Richardson, S. S., and Stevens, H. (Eds). (2015). *Postgenomics: Perspectives on Biology after the Genome*. Durham, NC: Duke University Press.

For the definition of the "postgenomic gene": Griffiths, P., and Stotz, K. (2013). *Genetics and Philosophy: An Introduction*. Cambridge: Cambridge University Press, p. 75.

On the definition of the postgenomic genome: Keller, E. F. (2015). The postgenomic genome. In Richardson, S. S., and Stevens, H. (Eds). *Postgenomics: Perspectives on Biology after the Genome*. Durham, NC: Duke University Press, 9–31.

Chapter 4

For the genetics of eye color: Sturm, R. A., and Frudakis, T. N. (2004). Eye colour: portals into pigmentation genes and ancestry. *TRENDS in Genetics*, 20(8), 327–332; Sturm, R. A., and Larsson, M. (2009). Genetics of human iris colour and patterns. *Pigment Cell & Melanoma Research*, 22(5), 544–562; White, D., & Rabago-Smith, M. (2011). Genotype–phenotype associations and human eye color. *Journal of human genetics, 56*(1), 5–7; Donnelly, M. P., Paschou, P.,

Grigorenko, E., Gurwitz, D., et al. (2012). A global view of the OCA2-HERC2 region and pigmentation. *Human Genetics, 131*(5), 683–696; Edwards, M., Cha, D., Krithika, S., et al. (2016). Iris pigmentation as a quantitative trait: variation in populations of European, East Asian and South Asian ancestry and association with candidate gene polymorphisms. *Pigment Cell & Melanoma Research*, 29(2), 141–162.

"80 percent of the variation in height is due to genetic factors": Visscher, P. M., Medland, S. E., Ferreira, M. A. R., et al. (2006). Assumption-free estimation of heritability from genome-wide identity-by-descent sharing between full siblings. *PLoS Genetics*, 2(3), e41.

Studies of the genetics of height: Visscher, P. M. (2008). Sizing up human height variation. *Nature Genetics*, 40(5), 489–490; Yang, J., Benyamin, B., McEvoy, B. P., et al. (2010). Common SNPs explain a large proportion of the heritability for human height. *Nature Genetics*, 42(7), 565–569; Yang, J., Manolio, T. A., Pasquale, L. R., et al. (2011). Genome partitioning of genetic variation for complex traits using common SNPs. *Nature Genetics*, 43(6), 519–525; Allen, H. L., Estrada, K., Lettre, G., et al. (2010). Hundreds of variants clustered in genomic loci and biological pathways affect human height. *Nature*, 467(7317), 832–838; Wood, A. R., Esko, T., Yang, J., et al. (2014). Defining the role of common variation in the genomic and biological architecture of adult human height. *Nature Genetics*, 46(11), 1173–1186; Yengo, L., Sidorenko, J., Kemper, K. E., et al. (2018). Meta-analysis of genome-wide association studies for height and body mass index in ~700000 individuals of European ancestry. *Human Molecular Genetics*, 27(20), 3641–3649.

MAOA studies: Brunner, H. G., Nelen, M. R., Van Zandvoort, P., et al. (1993). X-linked borderline mental retardation with prominent behavioral disturbance: phenotype, genetic localization, and evidence for disturbed monoamine metabolism. *American Journal of Human Genetics*, 52(6), 1032–1039; Brunner, H. G., Nelen, M., Breakefield, X. O., et al. (1993). Abnormal behavior associated with a point mutation in the structural gene for monoamine oxidase A. *Science*, 262(5133), 578–580; Caspi, A., McClay, J., Moffitt, T. E., et al. (2002). Role of genotype in the cycle of violence in maltreated children. *Science*, 297(5582), 851–854; Byrd, A. L., and Manuck, S. B. (2014). MAOA, childhood maltreatment, and antisocial behavior: meta-analysis of a gene–environment interaction. *Biological Psychiatry*, 75(1), 9–17; Ficks, C. A., and Waldman, I. D. (2014). Candidate genes for aggression and antisocial behavior: a meta-analysis of association studies of the 5HTTLPR and MAOAuVNTR. *Behavior Genetics*, 44(5), 427–444; Vassos, E., Collier, D. A., and Fazel, S. (2014). Systematic meta- analyses and field synopsis of genetic association studies of violence and aggression. *Molecular Psychiatry*, 19(4), 471–477; Odintsova, V. V., Roetman, P. J., Ip, H. F., et al. (2019). Genomics of human aggression: current state of genome-wide studies and an automated systematic review tool. *Psychiatric Genetics*, 29(5), 170–190.

On thalassemias: Weatherall, D. J. (2001). Phenotype–genotype relationships in monogenic disease: lessons from the thalassaemias. *Nature Reviews Genetics*, 2(4), 245–255; Higgs, D. R., Engel, J. D., and Stamatoyannopoulos, G. (2012). Thalassaemia. *The Lancet*, 379(9813), 373–383; Nuinoon, M., Makarasara, W., Mushiroda, T., et al. (2010). A genome-wide association identified the common genetic variants influence disease severity in β0-thalassemia/hemoglobin E. *Human Genetics*, 127(3), 303–314; Giardine, B., Borg, J., Viennas, E., et al. (2014). Updates of the HbVar database of human hemoglobin variants and thalassemia mutations. *Nucleic Acids Research*, 42(D1), D1063–D1069; Thein, S. L. (2013). The molecular basis of β- thalassemia. *Cold Spring Harbor Perspectives in Medicine*, 3(5), a011700.

On familial hypercholesterolemia: Goldstein, J. L., and Brown, M. S. (2015). A century of cholesterol and coronaries: from plaques to genes to statins. *Cell*, 161(1), 161–172; Usifo, E., Leigh, S. E. A., Whittall, R. A., et al. (2012). Low-density lipoprotein receptor gene familial hypercholesterolemia variant database: update and pathological assessment. *Annals of Human Genetics*, 76, 387–401; Abifadel, M., Varret, M., Rabs, J.-P., et al. (2003). Mutations in *PCSK9* cause autosomal dominant hypercholesterolemia. *Nature Genetics*, 34, 154–156; Zhang, D.-W., Lagace, T. A., Garuti, R., et al. (2007). Binding of proprotein convertase subtilisin/kexin type 9 to epidermal growth factor-like repeat A of low density lipoprotein receptor decreases receptor recycling and increases degradation. *Journal of Biological Chemistry*, 282, 18602–18612; Cohen, J. C., Boerwinkle, E., Mosley, T. H., Jr., and Hobbs, H. H. (2006). Sequence variations in PCSK9, low LDL, and protection against coronary heart disease. *New England Journal of Medicine*, 354, 1264–1272.

On the importance of mutations for cancer: Wu, S., Powers, S., Zhu, W., and Hannun, Y. A. (2016). Substantial contribution of extrinsic risk factors to cancer development. *Nature*, 529(7584), 43–47; Hanahan, D., and Weinberg, R. A. (2000). The hallmarks of cancer. *Cell*, 100(1), 57–70;

Hanahan, D., and Weinberg, R. A. (2011). Hallmarks of cancer: the next generation. *Cell*, 144(5), 646–674; Stratton, M. R., Campbell, P. J., and Futreal, P. A. (2009). The cancer genome. *Nature*, 458(7239), 719–724; Vogelstein, B., Papadopoulos, N., Velculescu, V. E., et al. (2013). Cancer genome landscapes. *Science*, 339(6127), 1546–1558; Tomasetti, C., and Vogelstein, B. (2015). Variation in cancer risk among tissues can be explained by the number of stem cell divisions. *Science*, 347(6217), 78–81.

On mutation rates in humans: Kong, A., Frigge, M. L., Masson, G., et al. (2012). Rate of de novo mutations and the importance of father's age to disease risk. *Nature*, 488(7412), 471–475; Sun, J. X., Helgason, A., Masson, G., et al. (2012). A direct characterization of human mutation based on microsatellites. *Nature Genetics*, 44(10), 1161–1165.

On the tissue organization field theory of cancer: Sonnenschein, C., and Soto, A. M. (2013). The aging of the 2000 and 2011 hallmarks of cancer reviews: a critique. *Journal of Biosciences*, 38(3), 651–663; Sonnenschein, C., and Soto, A. M. (2013). Cancer genes: the vestigial remains of a fallen theory. In Krimsky, S., and Gruber J. (Eds). *Genetic Explanations: Sense and Nonsense*. Cambridge, MA: Harvard University Press, 81–93; Sonnenschein, C., and Soto, A. M. (2020). Over a century of cancer research: inconvenient truths and promising leads. *PLoS Biology*, 18(4), e3000670.

Plutynski, A. (2018). *Explaining Cancer: Finding Order in Disorder*. Oxford: Oxford University Press; Plutynski, A. (2019). Cancer modelling: the advantages and limitations of multiple perspectives. In Massimi, M., and McCoy, C. D. (Eds). *Understanding Perspectivism: Scientific Challenges and Methodological Prospects*. New York: Routledge, 160–177.

Chapter 5

On the experiments of Roux and Driesch: Maienschein, J. (2014). *Embryos under the Microscope: The Diverging Meanings of Life*. Cambridge, MA: Harvard University Press, pp. 70–73.

Lewontin on metaphors of development, Lewontin, R. C. (2000). *The Triple Helix: Gene, Organism, and Environment*. Cambridge, MA: Harvard University Press, pp. 5, 6, 10–13.

On epigenesis and preformationism: Maienschein, J. (2017). Epigenesis and pre-formationism. *The Stanford Encyclopedia of Philosophy* (spring 2017 edition), Edward N. Zalta (Ed.), https://plato.stanford.edu/archives/spr2017/entries/epigenesis.

Alternative interpretation of the blueprint metaphor and related studies: Condit, C. M. (1999). *The Meanings of the Gene: Public Debates about Human Heredity*. Madison: University of Wisconsin Press, pp. 166–167; Parrott, R., and Smith, R. A. (2014). Defining genes using "blueprint" versus "instruction" metaphors: effects for genetic determinism, response efficacy, and perceived control. *Health Communication*, 29, 137–146.

For a clear and readable account of human development, on which my own account is also based: Davies, J. A. (2014). *Life Unfolding: How the Human Body Creates Itself*. Oxford: Oxford University Press, pp. 38–40, 192–193, 251–252.

On the origami metaphor: Wolpert, L. (2011). *Developmental Biology: A Very Short Introduction*. Oxford: Oxford University Press, p. 11.

On Treacher–Collins syndrome: Vincent, M., Geneviève, D., Ostertag, A., et al. (2016). Treacher Collins syndrome: a clinical and molecular study based on a large series of patients. *Genetics in Medicine*, 18(1), 49–56.

On *Achillea millefolium* plants: Núñez-Farfán, J., and Schlichting, C. D. (2001). Evolution in changing environments: the "synthetic" work of Clausen, Keck, and Hiesey. *Quarterly Review of Biology*, 76(4), 433–457.

On developmental plasticity and robustness: Bateson, P., and Gluckman, P. (2011). *Plasticity, Robustness, Development and Evolution*. Cambridge: Cambridge University Press.

On thigmomorphogenesis: Pigliucci, M. (2005). Evolution of phenotypic plasticity: where are we going now? *Trends in Ecology & Evolution*, 20(9), 481–486.

On the *SRY* gene: Jäger, R. J., Anvret, M., Hall, K., and Scherer, G. (1990). A human XY female with a frameshift mutation in the candidate testis-determining gene SRY. *Nature*, 348, 452–454; Margarit, E., Coll, M. D., Oliva, R., et al. (2000). SRY gene transferred to the long arm of the X chromosome in a Y-positive XX true hermaphrodite. *American Journal of Medical Genetics*, 90(1), 25–28;

Sekido, R., & Lovell-Badge, R. (2009). Sex determination and SRY: down to a wink and a nudge? *Trends in Genetics*, 25(1), 19–29.

On Francis Galton's research on twins: Burbridge, D. (2001). Francis Galton on twins, heredity and social class. *The British Journal for the History of Science*, 34(3), 323–340.

On the Jensen-Lewontin debate: Jensen, A. R. (1969). How much can we boost IQ and scholastic achievement? *Harvard Educational Review*, 39, 1–123; Lewontin, R. C. (1970). Race and intelligence. *Bulletin of Atomic Scientists*, 26(3), 2–8; Tabery, J. (2014). *Beyond Versus: The Struggle to Understand the Interaction of Nature and Nurture*. Cambridge, MA: MIT Press.

Cattle example of heritability: Slack, J. (2014). *Genes: A Very Short Introduction*. Oxford: Oxford University Press, p. 75.

Men laying bricks example of heritability: Lewontin, R. C. (1974). The analysis of variance and the analysis of causes. *American Journal of Human Genetics*, 26, 400–411.

Snowfall example of heritability: Moore, D. S. (2002). *The Dependent Gene: The Fallacy of "Nature vs. Nurture."* New York: Times Books/Henry Holt and Company, p. 41.

Drummers example: Keller, E. F. (2010). *The Mirage of a Space between Nature and Nurture*. Durham, NC: Duke University Press, pp. 34–36.

On the concept of difference-maker: Waters, C. K. (2007). Causes that make a difference. *Journal of Philosophy*, 104, 551–579.

On lactose intolerance: Gannett, L. (1999). What's in a cause? The pragmatic dimensions of genetic explanations. *Biology and Philosophy*, 14, 349–373.

Chapter 6

For examples of the many-to-many relation: Frazer, K. A., Murray, S. S., Schork, N. J., and Topol, E. J. (2009). Human genetic variation and its contribution to complex traits. *Nature Reviews Genetics*, 10(4), 241–251.

On knockout genes in mice: Morange, M. (2002). *The Misunderstood Gene*. Cambridge MA: Harvard University Press; Hall, B., Limaye, A., and Kulkarni, A. B. (2009). Overview: generation of gene knockout mice. *Current Protocols in Cell Biology*, Unit 19– 12, 11–17; Mak, T. W., Penninger, J. M., and Ohashi, P. S. (2001). Knockout mice: a paradigm shift in modern immunology. *Nature Reviews Immunology*, 1(1), 11–19.

On human knockouts: Alkuraya, F. S. (2015). Human knockout research: new horizons and opportunities. *Trends in Genetics*, 31(2), 108–115; Shaheen, R., Faqeih, E., Ansari, S., et al. (2014). Genomic analysis of primordial dwarfism reveals novel disease genes. *Genome Research*, 24(2), 291–299; MacArthur, D. G., Balasubramanian, S., Frankish, A., et al. (2012). A systematic survey of loss-of-function variants in human protein-coding genes. *Science*, 335, 823–828; Sulem, P., Helgason, H., Oddson, A., et al. (2015). Identification of a large set of rare complete human knockouts. *Nature Genetics*, 47(5), 448–452; Chen, R., Shi, L., Hakenberg, J., et al. (2016). Analysis of 589,306 genomes identifies individuals resilient to severe Mendelian childhood diseases. *Nature Biotechnology*, 34(5), 531–538.

On gene pleiotropy: Solovieff, N., Cotsapas, C., Lee, P. H., Purcell, S. M., and Smoller, J. W. (2013). Pleiotropy in complex traits: challenges and strategies. *Nature Reviews Genetics*, 14(7), 483–495.

On epigenetics: Moore, D. S. (2015). *The Developing Genome: An Introduction to Behavioral Epigenetics*. Oxford: Oxford University Press; Carey, N. (2012). *The Epigenetics Revolution: How Modern Biology Is Rewriting Our Understanding of Genetics, Disease, and Inheritance*. New York: Columbia University Press; Francis, R. C. (2011). *Epigenetics: How Environment Shapes our Genes*. New York: W.W. Norton.

Study with twins: Fraga, M. F., Ballestar, E., Paz, M. F., et al. (2005). Epigenetic differences arise during the lifetime of monozygotic twins. *Proceedings of the National Academy of Sciences*, 102(30), 10604–10609.

On the Dutch famine: de Rooij, S., Wouters, H., Yonker, J., et al. (2010). Prenatal undernutrition and cognitive function in late adulthood. *Proceedings of the National Academy of Sciences*, 107, 16881–16886; Roseboom, T., de Rooij, S., and Painter, R. (2006). The Dutch Famine and its long-term consequences for adult health. *Early Human Development*, 82(8), 485–491; Heijmans, B., Tobi, E., Stein, A., et al. (2008). Persistent epigenetic differences associated with prenatal exposure to famine in humans. *Proceedings of the National Academy of Sciences*, 105(44), 17046–17049; Tobi, E., Goeman, J.,

Monajemi, R., et al. (2014). DNA methylation signatures link prenatal famine exposure to growth and metabolism. *Nature Communications*, 5, 5592.

Quotations from Collins: Collins, F. S. (2010). *The Language of Life: DNA and the Revolution in Personalized Medicine*. New York: Harper, pp. xxiii–xxiv, 278–279.

On genomic medicine and the related questions: Annas, G. J., and Elias, S. (2015). *Genomic Messages: How the Evolving Science of Genetics Affects Our Health, Families, and Future*. New York: HarperOne; Collins, F. S., and Varmus, H. (2015). A new initiative on precision medicine. *New England Journal of Medicine*, 372, 793–795; Lander, E. S. (2015). Cutting the Gordian helix: regulating genomic testing in the era of precision medicine. *New England Journal of Medicine*, 372(13), 1185–1186.

Estimation of relative risk for five individuals: Ng, P. C., Murray, S. S., Levy, S., and Venter, J. C. (2009). An agenda for personalized medicine. *Nature*, 461(7265), 724–726.

Comparison of 23andMe, deCODEme, and Navigenics: Kalf, R. R., Mihaescu, R., Kundu, S., et al. (2014). Variations in predicted risks in personal genome testing for common complex diseases. *Genetics in Medicine*, 16(1), 85–91.

Poll among experts: Schulte, J., Rothaus, C. S., Adler, J. N., and Phimister, E. G. (2014). Screening an asymptomatic person for genetic risk: polling results. *New England Journal of Medicine*, 371(20), 2442–2445.

For the 2013 decision of the FDA regarding 23andMe: Annas, G. J., and Elias, S. (2014). 23andMe and the FDA. *New England Journal of Medicine*, 370(11), 985–988; Green, R. C., and Farahany, N. A. (2014). The FDA is overcautious on consumer genomics. *Nature*, 505, 286–287.

For the 2018 decision of the FDA regarding 23andMe: Gill, J., Obley, A. J., and Prasad, V. (2018). Direct-to-consumer genetic testing: the implications of the US FDA's first marketing authorization for BRCA mutation testing. *Journal of the American Medical Association*, 319(23), 2377–2378.

Study of raw DNA data provided by companies performing genetic tests: Tandy-Connor, S., Guiltinan, J., Krempely, K., et al. (2018). False-positive results released by direct-to-consumer genetic tests highlight the importance of clinical confirmation testing for appropriate patient care. *Genetics in Medicine*, 20(12), 1515.

On polygenic risk scores: Torkamani, A., Wineinger, N. E., and Topol, E. J. (2018). The personal and clinical utility of polygenic risk scores. *Nature Reviews Genetics*, 19(9), 581–590; Sugrue, L. P., and Desikan, R. S. (2019). What are polygenic scores and why are they important? *Journal of the American Medical Association*, 321(18), 1820–1821.

GeneScreen study: Butterfield, R. M., Evans, J. P., Rini, C., et al. (2019). Returning negative results to individuals in a genomic screening program: lessons learned. *Genetics in Medicine*, 21(2), 409–416.

On the role of metaphors in science: Lakoff, G., and Johnson, M. (2003/1980). *Metaphors we Live By*. Chicago: University of Chicago Press; Brown, T. L. (2003). *Making Truth: Metaphor in Science*. Urbana and Chicago: University of Illinois Press; Reynolds, A. S. (2018). *The Third Lens: Metaphor and the Creation of Modern Cell Biology*. Chicago: University of Chicago Press; Keller, E. F. (1995). *Refiguring Life: Metaphors of Twentieth Century Biology*. New York: Columbia University Press.

Sources of quotations: Avise, J. C. (2001). Evolving genomic metaphors: a new look at the language of DNA. Science, 294(5540), 86–87, p. 86; Pollack, R. (1995). *Signs of Life: The Language and Meanings of DNA*. London: Penguin, p. 90.

On gene editing and CRISPR: Lander, E. S. (2016). The heroes of CRISPR. *Cell*, 164 (1–2), 18–28; Kobuzek, J. (2016). *Modern Prometheus: Editing the Human Genome with CRISPR-CAS9*. Cambridge: Cambridge University Press; Parrington, J. (2016). *Redesigning Life: How Genome Editing Will Transform the World*. Oxford: Oxford University Press; Doudna, J., and Sternberng, S. (2017). *A Crack in Creation: The New Power to Control Evolution*. London: Vintage (quotations on pp. xv, xvii); Carey, N. (2019). *Hacking the Code of Life: How Gene Editing Will Rewrite our Futures*. London: Icon Books.

On the "gene editing" and "book of life" metaphors: Nerlich, B. "The book of life: reading, writing and editing," https://blogs.nottingham.ac.uk/makingscience public/2015/11/22/the-book-of-life-reading-writing-and-editing (accessed June 29, 2020); Nerlich, B., Dingwall, R., and Clarke, D. D. (2002). The book of life: How the completion of the Human Genome Project was revealed to the public. *Health*, 6(4), 445–469; Ceccarelli, L. (2004). Neither confusing cacophony nor culinary complements: a case study of mixed metaphors for genomic science. *Written Communication*, 21(1), 92–105.

Study on "editing" and "targeting" metaphors in articles: O'Keefe, M., Perrault, S., Halpern, J., et al. (2015). "Editing" genes: a case study about how language matters in bioethics. *The American Journal of Bioethics*, 15(12), 3–10.

Report: National Academy of Sciences (2020). *Heritable Human Genome Editing*. Washington, DC: National Academies Press.

Studies on problems of genome editing: www.nature.com/articles/d41586-020-019 06-4?utm_source=Nature+Briefing&utm_campaign=6441803628-briefing-dy-20200626&utm_medium=email&utm_term=0_c9dfd39373-6441803628-42984447 (accessed June 26, 2020).

For a review of genome editing methods: Anzalone, A. V., Koblan, L. W., and Liu, D. R. (2020). Genome editing with CRISPR–Cas nucleases, base editors, transposases and prime editors. *Nature Biotechnology*, 38, 824–844.

On genomic prediction: https://genomicprediction.com/faqs/#faq-7.2 (accessed June 29, 2020).

Interview with Stephen Hsu: www.genengnews.com/insights/polygenic-risk-scores-and-genomic-prediction-qa-with-stephen-hsu (accessed June 29, 2020).

For the criticisms toward the initial announcement of Genomic Prediction: www .technologyreview.com/2019/11/08/132018/polygenic-score-ivf-embryo-dna-tests-genomic-prediction-gattaca (accessed June 29, 2020); http://ewan birney.com/2019/11/why-using-genetic-risk-scores-on-embryos-is-wrong .html (accessed June 29, 2020).

Study of PRS of height and IQ: Karavani, E., Zuk, O., Zeevi, D., et al. (2019). Screening human embryos for polygenic traits has limited utility. *Cell*, 179(6), 1424–1435.

Confusing "the methodological limitations of experiments" with "the correct explanations of phenomena": Lewontin, R. C. (2000). *The Triple Helix: Gene, Organism, and Environment*. Cambridge, MA: Harvard University Press, pp. 98–99.

A recent analysis of the literature on geneticization: Weiner, K., Martin, P., Richards, M., and Tutton, R. (2017). Have we seen the geneticisation of society? Expectations and evidence. *Sociology of Health & Illness*, 39(7), 989–1004.

Index

Locators in **bold** refer to tables; those in *italic* to figures; numbers are filed as spelt.

absolute risk, public understanding, 23, **24**
Achillea millefolium (yarrow), characteristics, 131–132, 142
age-related macular degeneration, 165–166, **167**, **168**
aggression, genetics of, 101–105, *104*
allele(s), 40, 88
allelomorphs, origin of concept, 40
alternative RNA splicing, 66–68, *69*
Angelman syndrome, 158–159
animal studies, functional products of genes, 85–86; *see also* bacteria; cat coat markings; cattle; *Drosophila*; frogs; horses; mice; sea urchins
Arabidopsis thaliana (thale cress), characteristics, 133
associated genes, 94, 162–172, 188; *see also* genome wide association studies
atrial fibrillation, 165–166, **167**, **168**
autoimmune diseases, 149
Avery, Oswald, 49
Avise, John, 173

bacteria, transgenic DNA technologies, 176–177
Baltimore, David, 72
Bateson, William, 38–40
BCL11A (B-cell CLL/lymphoma 11A) gene, 107
Beadle, George, 48
beads on a string model, chromosomes, 43, *44*, 54
behavioral characteristics, aggression, 101–105, *104*
behavioral genetics, public understanding, 14–16
Benzer, Seymour, 54
biological characteristics *see* characteristics
biological specificity concept, 55
blueprint metaphor, development, 122–124, *123*, 127
book of life metaphor, 178–179, 188
BRCA1 and *BRCA2* (BReastCAncer) genes, 150
    human knockout studies, 151–152
    risk and probability of disease, 23–30, 172

breast cancer, 150
    Angelina Jolie's mastectomy, 17–22, 23, 172
    epigenetics, 160
    risk and probability of disease, 23–30, **26**
Brenner, Sydney, 56
brick wall example, gene–environment interactions, 143–144
Brno Natural Science Society, 35, 37
Brooks, William Keith, 37
Bull, Sofia, 10

calico cat coat marking, 155–157
cancers, genetics of, 113–120; *see also* breast cancer
Carver, Rebecca Bruu, 6
Cas (CRISPR-associated) protein 9, 177–178, 179–180
Castle, William, 40
cat coat markings, 155–157
cattle example, heritability, 143
causation
    and association, 89, 162–172
    misunderstandings, 188
CDK6 (cyclin-dependent kinase 6) gene, 99–101
Celera Genomics, 76
celiac disease, 165–166, **167**, **168**
cell signaling/interactions, 126–127, 130, 173–174
central dogma
    Francis Crick, 55, 59, *60*
    postgenomics era, 161, *162*
characteristics, genetics of; *see also* development of characteristics; gene–characteristic relationship
    aggression, 101–105, *104*
    brick wall example, 143–144
    eye color, 96–101, *97, 99*

gene–environment interactions, 138–145
genes as characteristic-makers, 43
genes as difference-makers, 43, 145–148
genetic vs. environmental contributions, 14, **15**, 16–17
    height, 99–101, 132
    heritability, 142–145
    IQ, 139–141
    Mendel's work on peas, 31–41, *32, 34*
    monogenic diseases, 105, *114–115*
    multifactorial diseases/cancers, 113–120
    relatives and twin studies, 138–139, *140*
    variation within populations, 138–148, *140*
Charpentier, Emmanuelle, 177–178
Chase, Martha, 49
chromatin, 79, *80*, 153
chromosomes, 43, *44*
classical genetics era, xviii–xix, 41–48, *44, 46*, 61
clinical utility, disease risk in individuals, 172
codons, 56
Collins, Francis, 73, 76, *77*, 162, 164–165
common misunderstandings *see* public understanding of genetics
conceptual familiarity, public understanding, 11–12, *13*
Condit, Celeste, 6, 123
control genes, origins of concept, 57–64, *62*
Coriell Personalized Medicine Collaborative study, 12–14
coronary heart disease, 153
    disease risk in individuals, 165
    epigenetics, 160

Correns, Carl, 38
"credit card gene", 14–16
Creighton, Harriet, 47
Crick, Francis
    central dogma, 55, 59, *60*
    sequence hypothesis, 54–55
    structure of DNA, 49–53, *51*
CRISPR (Clustered Regularly Interspaced
    Short Palindromic Repeats), gene
    editing, 177–178, 179–180
Crohn's disease, 149, 153, 165–166,
    167, **168**

Darwin, Charles, 36
Davies, Kevin, 164–165
de Vries, Hugo, 37, 38
deCODEme test, 164–166, **167**, **168**
decoding, public understanding, 188
deficit model, expert vs. public
    understanding, 1–2
descriptive plans, developmental,
    127–128, *128*
determinism, genetic, xvii
    developmental, 131, 132
    gene concept, 6, **8**
    media representations of genetics,
        5–7, **7**
    misunderstandings, 187–188
development, definition, 122
development of characteristics,
    121–130, *126*
    Achillea millefolium example,
        131–132, 142
    Arabidopsis thaliana example, 133
    blueprint metaphor, 122–124,
        123, 127
    cell signaling/interactions,
        126–127, 130
    descriptive plans, 127–128, *128*
    environmental influences, 121–122,
        130–138, *135*

generative plans, 127–128, *128*
holistic explanations, 133–134
instruction metaphor, 124
levels of organization, *134*
misrepresentations of, *135*
origami metaphor, 127
Phantom of the Opera example,
    128–130, 132
relatives and twin studies,
    138–139, *140*
SRY gene, 135–137, *136*
TCOF1 gene, 137
variation within populations,
    138–148, *140*
developmental genes, origins of concept,
    61–64
developmental plasticity, 131–132
developmental robustness, 131, 132
diabetes, type 1, 153
diabetes, type 2, 149, 165–166, **167**, **168**
Dickinson, Boonsri, 164–165
difference-makers, genes as, 14, **15**,
    16–17, 43, 145–148
differentiation process,
    development, 125
direct-to-consumer genetics (DTCG),
    164–165
disease; *see also specific diseases*
    genetic, definition, 118
    monogenic, 105, *114–115*
    multifactorial, 113–120
    risks *see* risk and probability of disease
dissemination of science, 2
DNA (deoxyribonucleic acid)
    double helix model, 49–53, *51*
    methylation, 154–155, *156*
    origin of concept, 49
    relationship to genes, 187
    replication, 50–53, *52–53*
    testing *see* genetic testing
double helix model, DNA, 49–53, *51*

double mastectomy *see* Jolie, Angelina
Doudna, Jennifer, 177–178
Dr. Jekyll and Mr. Hyde film, 7
Driesch, Hans, 121–122
driver mutations, cancer, 115–116
Drosophila (fruit flies)
    Morgan's work, 41–48
    X-linked inheritance, 45, *46*
drum/drumming example, variation, 145
DTCG (direct-to-consumer genetics),
    164–165
Dutch famine, 159–160

ENCODE project (Encyclopedia of DNA
    Elements), xix, 79–86, 153, 173
environmental influences; *see also*
    epigenetics; gene–environment
    interactions
    developmental, 121–122, 130–138,
    134, 135
    individual differences, 14, **15**, 16–17
epigenetics, *156*, 153–162
    central dogma, 161, *162*
    DNA methylation, 154–155, *156*
    genomic/epigenetic imprinting,
    157–159
    histone modification, 155, *156*
essentialism, genetic, xvii, 5–6, 16–17,
    187
evolutionary frameworks, gene concept,
    6, **8**
exceptionalism, genetic, 12, *13*
exome sequencing, 75
exon, RNA splicing, 65–72, *69*
expert vs. public understanding of
    genetics, 1–2
exploded gene concept, 72
expressivity, 150; *see also* gene
    expression
eye color, 96–101, *97, 99*
    environmental influences, 132

variation within populations, 145

familial hypercholesterolemia, 111–112,
    114–115
famine, Dutch, 159–160
fantasy, genetic, 9
fatalism, genetic, xvii–xviii, 184, 186
film portrayals of genetics, 7–10
"financial success gene", 3
5HTTLPR (serotonin-transporter-linked
    polymorphic region), serotonin
    transporter gene, 104
Fleming, Nic, 164–165
Flemming, Walther, 38
forest fire example, genes as difference-
    makers, 146
Frankenstein film, 7
frogs, development, 121
functional genome, 81–86, *82*, 187

Galton, Francis, 36–37, 38, 138
Gamow, George, 55
Gärtner, Friedrich von, 35
GATTACA (dystopian film), xvi,
    xvii, 182
gene(s); *see also* characteristics (genetics
    of); origin of gene concept; public
    understanding of genetics
    beads on a string model, 43, *44*, 54
    as characteristic-makers, 43
    definitions, 5–6, 81, *82*, 186
    as difference-makers, 14, **15**,
    16–17, 43, 145–148
    knockout, 150–152
    media representations, 6, **8**
    metaphors, 172–181, 188
    misunderstandings about, 188
    origin of term, 40–41, 44
    relationship to DNA, 187
    relativistic frameworks, 6, **8**
gene action metaphor, 174–175

gene–characteristic relationship, complexity of, 149–153
    disease risk in individuals, 162–172, **167**, **168**
    epigenetics, 153–162, *156*
    gene metaphors, 172–181
    genetic tests, 162–172, **167**, **168**
    misunderstandings, 188
    postgenomics era, 161, *162*
gene editing metaphor, 175–179
gene–environment interactions; *see also* environmental influences; epigenetics
    brick wall example, 143–144
    variation within populations, 138–145
gene expression, 56–57, *58*
    regulatory/control genes, 57, *62*
    RNA splicing, *67*
gene interaction metaphors, 174–175
generative plans, developmental, 127–128, *128*
genetic determinism *see* determinism
genetic disease, definition, 118
genetic essentialism, xvii, 5–6, 16–17, 187
genetic exceptionalism, 12, *13*
genetic fatalism, xvii–xviii, 184, 186
genetic inheritance, definition, 36
genetic realism, 9
genetic reductionism, xvii, 188
genetic testing, xvi, 182–183
    brand comparisons, 165–166, **167**, **168**
    disease risk in individuals, 162–172
    participant reactions to/understanding of, 171–172
    polygenic risk scores, 170–171
geneticization, 4–5, 184
genome wide association studies (GWAS), xix, 86–95
    gene–characteristic relationship, 149

ice cream and water consumption example, 89, *92*
    linkage disequilibrium, *90–91*
    pleiotropy, 152–153
    single nucleotide polymorphisms, 86–95
    single nucleotide variants, 86–88, *87*
genomes, misunderstandings about, 188
genomic imprinting, 157–159
genomic medicine, 163
Genomic Prediction (genetic testing company), 182–183
genomics, xix
genotype, origin of concept, 40–41
germline, 38
Gigerenzer, Gerd, 23
Gilbert, Walter, 73
glucose tolerance, epigenetics, 160
Gros, François, 72
growth processes, 125; *see also* development of characteristics
GWAS *see* genome wide association studies

Hamner, Everett, 9
Hanahan, Douglas, 117
hard heredity, 38
HBB (hemoglobin subunit beta) gene, 150, 183
    genes as difference-makers, 146–147
    thalassemia, 106–110, *108*, *110*
HBS1 L (HBS1 like translational GTPase) gene, 107
height
    environmental influences, 132
    genetics of, 99–101
Heine, Steven, 16–17
hemoglobin subunit beta gene *see HBB*
HERC2 (hect domain and RCC1-like domain-containing protein 2) gene, 98–99

heredity, Mendel's work, 36–38

heritability, 140, 142–145

Heritable Human Genome Editing (International Commission on the Clinical Use of Human Germline Genome Editing), 180

heritable human genome editing (HHGE), 180

Hershey, Alfred, 49

Hertwig, Oscar, 38

heterozygote, origin of concept, 40

HHIP (hedgehog interacting protein) gene, 99–101

Hieracium (Hawkweed spp.), Mendel's work, 37

hinnies, genomic/epigenetic imprinting, 158

histone modification, epigenetics, 155, *156*

historical perspectives, xviii–xix; *see also* origin of gene concept

HMGA2 (high-mobility group AT-hook 2) gene, 99–101

holistic explanations, development of characteristics, 133–134

homozygote, origin of concept, 40

horses, genomic/epigenetic imprinting, 158

Hsu, Stephen, 182

human embryo development, *126*; *see also* development of characteristics

Human Genome Project (HGP), 72–79, *77*

human knockout studies, 151–152

human-focused perspective of the text, xviii

hybridization

   Mendel's work, 36

   plant breeding, 33

ice cream and water example, GWAS, 89–92

identical twins *see* twin studies

IGF2 gene (insulin like growth factor 2) gene, 160

imprinting, genomic/epigenetic, 157–159

independent assortment, Mendel's law, 33

individual differences *see* characteristics; risk and probability of disease

induced pluripotent stem (iPS) cells, 161

"infidelity gene", 3–4

inheritance

   definition, 36

   X-linked inheritance, 45, *46*

inherited disease, definition, 118

instruction metaphor of development, 124

interactions of genes with environment *see* gene–environment interactions

International Commission on the Clinical Use of Human Germline Genome Editing, 180

interspecific comparisons *see* animal studies

intron sequences, RNA splicing, 65–72, *69*

IQ (intelligence quotient), 14, **15**, 139–141

Jacob, François, 55, 57

JAZF1 (JAZF1 zinc finger 1) gene, 149

Jensen, Arthur, 139–141

Johannsen, Wilhelm, 40–41

Jolie, Angelina

   media representations, 17–22

   risk and probability of disease, 23, 172

junk DNA, 77–79, 83

Keller, Evelyn Fox, 73–74, 145
Kirby, David, 7
knockout gene studies, 150–152
Kölreuter, Josef Gottlieb, 35
Kornberg, Roger, 79

lactose intolerance example, 147
Lander, Eric, 74, 76, 163–164
law of independent assortment,
     Mendel's, 33
law of segregation, Mendel's, 33
LDL (low density lipoprotein) levels,
     153, 165
LDL-R (low-density lipoprotein receptor)
     gene, 111, *114–115*, 150
Lewontin, Richard, 122, 141–142,
     184
"liberal gene", 14–16
Lindee, Susan, 5–6
linkage disequilibrium, *90–91*
Linnaeus, Carl, 35
lipid profiles, epigenetics, 160; *see
     also* LDL
Lippman, Abby, 4–5, 184
Lmbr1 (limb development membrane
     protein 1), 61–64
Lmbr1 gene, *63*

McCarty, Maclyn, 49
McClintock, Barbara, 47
MacLeod, Colin, 49
macular degeneration, 165–166,
     167, **168**
many genes–one characteristic model,
     93, 188; *see also*
     gene–characteristic relationship
mapping, genetic, 44, *44*
mastectomy, double *see* Jolie, Angelina
materialistic frameworks, gene concept,
     6, **8**
Matthaei, Heinrich, 56

media representations of genetics, 1–10;
     see also public understanding of
     genetics
   Angelina Jolie's double mastectomy,
        17–22
   film portrayals, 7–10
   gene frameworks, 6, **8**
   genetic determinism, 5–7, **7**
   genetic essentialism, 5–6
   public image of genes, 1–10
   TV portrayals, 10
meiosis, 38, *39*
Mendel, Gregor
   Mendelian inheritance, 31–43, *32*, *34*
   law of independent assortment, 33
   law of segregation, 33
Mendel's Principles of Heredity (Bateson),
     38–40
messenger RNA (mRNA), 55
meta-fiction, genetic, 9
methodological limitations, genetic
     experiments, 184, 185, 186
methylation, DNA, 154–155, *156*
mice
   gene knockout, 151
   genomic/epigenetic imprinting, 157
misunderstandings about genes *see*
     public understanding
mitosis, 38, *39*
molecular genetics era, xviii–xix, 48–57,
     52–53, *58*, 61
monoamine oxidase A (MAOA) enzyme,
     101–105, *104*
Monod, Jacques, 55, 57
monogenic diseases, 105, *114–115*
Moore, David, 144
Morgan, Thomas Hunt, 41–48, 145, 185
mouse studies *see* mice
mules, genomic/epigenetic
     imprinting, 158
Muller, Hermann J. 47

multifactorial diseases, genetics of, 113–120
multipotent cells, epigenetics, 161
multipotent embryonic stem cells, 151
mutations, 53–54
    cancers, 113–120
    monogenic diseases, 105, *114–115*
MYB (MYB proto-oncogene) gene, 107

Nägeli, Carl von, 37
Napp, Cyril, 33
nature–nurture debate, 138, 148; *see also*
        environmental influences;
        gene–environment interactions
Navigenics test, 164–166, **167**, **168**
negative predictive value, genetic testing, 169–170
Nelkin, Dorothy, 5–6
Netherlands famine, 159–160
Neurospora crassa (bread mold), 48–49
Nirenberg, Marshall, 56
nomenclature, genes and proteins, xviii

obesity, genetic and environmental contributions, **15**, 16–17
obstructive airways disease, epigenetics, 160
OCA2 gene (OCA2 melanosomal transmembrane protein), 98–99
one gene–many characteristics model, 93
one gene–one characteristic model, 72, 77, 120, 183; *see also*
        gene–characteristic relationship
one gene–one enzyme model, 48–49
origami metaphor, development of characteristics, 127
origin of gene concept
    alternative splicing, 66–68, *69*
    associated genes, 94
    beads on string model, 43, *44*, 54
    central dogma, 55, 59, *60*

chromatin, 79, *80*
classical genetics era, 41–48, *44, 46*, 61
developmental genes, 61–64
ENCODE project, 79–86
exon/intron sequences, 65–72, *67, 69*
exploded gene concept, 72
functional products of genes, 81–86, *82*
GWAS, 86–95, *87, 92*
Human Genome Project, 72–79, *77*
junk DNA, 77–79, 83
linkage disequilibrium, *90–91*
Lmbr1 gene, 61–64, *63*
many genes–one characteristic model, 93
Mendel's work, 31–41, *32, 34*
mitosis and meiosis, *39*
molecular genetics era, 48–57, *52–53, 58*, 61
one gene–many characteristics model, 93
overlapping genes, 70, *71*
postgenomics, 94–95
regulatory/control genes, 57–64, *62*
RNA splicing, 65–72, *67, 69*
single nucleotide polymorphisms, 86–95
single nucleotide variants, 86–88, *87*
transcription factors, *60*
trans-splicing, 69–70, *70*
X-linked inheritance, 45, *46*
O'Riordan, Kate, 2
ovarian cancer, 17–22
overlapping genes, 70, *71*, 187

pangenesis, 36, 40–41
passenger mutations, cancer, 115–116
Pauling, Linus, 49, 54
PCSK9 (proprotein convertase subtilisin/ kexin type 9) gene, 112, 150

peas, Mendel's work, 31–41, *32, 34*

penetrance, 150

personalized medicine, 163, 172

Phantom of the Opera example
  development of characteristics, 128–130, 132
  epigenetics, 154–155

phenotype, origin of concept, 40–41

Pisum sativum (pea), Mendel's work, 31–41, *32, 34*

plants; *see also Achillea millefolium; Arabidopsis thaliana; Hieracium*
  hybridization, 33
  Mendel's work on peas, 31–41, *32, 34*
  speciation, 35

plasticity, developmental, 131–132

pleiotropy, 152–153

pluripotent cells, epigenetics, 161

Plutynski, Anya, 119

Pollack, Robert, 175

polygenic risk scores (PRSs), 170–171, 182

positive predictive value, genetic testing, 169–170

postgenomics era, xix, 94–95, 161, *162*

potency, cell, 161

Prader–Willi syndrome, 158–159

Precision Medicine Initiative, USA, 164

preformationism theories, developmental, 122–124

pre-genetics era, xviii

Principles of Biology (Spencer), 36

probability of disease *see* risk and probability of disease

prostate cancer, 149, 165–166, **167, 168**

proteins, xviii, 183

proto-oncogenes, cancers, 115

Provisional Hypothesis of Pangenesis (Darwin), 36

PTPN22 (protein tyrosine phosphatase, nonreceptor type 22) gene, 149, 153

public understanding of genetics, 11–17; *see also* media representations
  Angelina Jolie's double mastectomy, 17–22
  behavioral genetics, 14–16
  common misunderstandings, 187–188
  vs. expert understanding, 1–2
  conceptual familiarity, 11–12, *13*
  genetic essentialism, 16–17
  individual differences, 14, **15**
  public perceptions of importance, 185–186
  risk and probability of disease, 23–30, **24, 26**

reaction norms, developmental, 131

realism, genetic, 9

reductionism, genetic, xvii, 188

regulatory genes, origins of concept, 57–64, *62*

relative risk, public understanding, 23

relativistic frameworks, gene concept, 6, **8**

rheumatoid arthritis, 153

risk and probability of disease, 23–30, 170–171
  absolute and relative risk, **24**
  complexity of, 162–172, **167, 168**, 188
  polygenic risk scores, 170–171
  public understanding, 23–30, **26**, 172

RNA splicing, 65–72, *67*

robustness, developmental, 131, 132

rodent studies *see* mice

role of genes *see* characteristics (genetics of)

Roux, Wilhelm, 121–122

science, mediation/dissemination of, 2

science fiction, 7–10, 185

screening *see* genetic testing

sea urchins, development, 121–122
segregation, Mendel's law, 33
sensitivity, genetic testing, 169–170
sequence hypothesis, Crick, 54–55
serotonin-transporter-linked
    polymorphic region (5HTTLPR),
    serotonin transporter gene,
    104
sickle cell anemia, 54
single nucleotide polymorphisms (SNPs)
    disease risk in individuals,
        165–166, **167**
    gene–characteristic relationship, 149
    GWAS, 86–95
single nucleotide variants (SNVs),
    86–88, *87*
singular gene paradigm, 30
snowflakes, formation, 144
SNPs *see* single nucleotide
    polymorphisms
"social media use gene", 3
somatic mutation theory (SMT) of
    cancer, 113
somatic mutations, 118
Sonnenschein, Carlos, 118
Soto, Ana, 118
SOX9 (sex determining region Y-box 9)
    gene, 136–137
speciation of plants, 35
specificity, genetic testing, 169–170
Spencer, Herbert, 36
splicing, RNA, 65–72, *67*
SRY (sex-determining region Y) gene,
    135–137, *136*, 146
"stardom gene", 3
Steinberg, Deborah Lynn, 30
stress responsiveness, epigenetics, 160
structural genes, origins of concept,
    57–58
Sturtevant, Alfred, 44
Sutton, Walter, 40

symbolic frameworks, gene concept, 6, **8**
systemic lupus erythematosus, 153

Tabery, James, 142
Tatum, Edward, 48
TCOF1 (treacle ribosome biogenesis
    factor 1) gene, 137
testing *see* genetic testing
thalassemia, 105–113, *110*, 146–147,
    150, 183
The Theory of the Gene (Morgan), 45
tissue organization field theory (TOFT),
    cancer, 118–119
totipotent cells, epigenetics, 161
transcription factors, 59, *60*
transgenic DNA technologies, 176–177
trans-splicing, RNA splicing, 69–70, *70*
Treacher-Collins syndrome, 137
TV portrayals of genes/genetics, 10
23andMe test, 164–169, **167**, **168**
twin studies, 138–139, *140*, 159
type 1 diabetes, 153
type 2 diabetes, 149, 165–166,
    167, **168**

UBE3A (ubiquitin protein ligase E3A)
    gene, 159
Unger, Franz, 35
unipotent cells, epigenetics, 161
unit characters, 41–48
unit factors, 41

Van Dijck, J. 2
variation within populations,
    characteristics, 145–148
Venter, Greg, 76, *77*, 173
Vogelstein, Bert, 116
von Tschermak, Erich Seysenegg, 38

water consumption example, GWAS,
    89–92

Watson, James, structure of DNA,
    49–53, *51*
Weinberg, Robert A. 117
WEIRD (Western, Educated,
    Industrialized, Rich, and
    Democratic), 11
Weismann, August, 37, 38
Weldon, Raphael, 40

what genes do *see* characteristics
    (genetics of)
whole-genome sequencing, 74

X-linked inheritance, 45, *46*

ZBTB38 (zinc finger and BTB domain
    containing 38) gene, 99–101